Conference Organizing Committee

IMechE Conference Transactions

I MECH E

International Conference on

Fault-Free Infrastructure

Effective Solutions to Improve Efficiency

23–24 November 1999
Pride Park Stadium, Derby, UK

Organized by
Railway Division of the Institution of
Mechanical Engineers (IMechE)

Sponsored by
The Institution of Electrical Engineers (IEE)
The Institution of Railway Signal Engineers

IMechE Conference Transactions 1999–11

Professional Engineering Publishing

Published by Professional Engineering Publishing Limited for The Institution of Mechanical Engineers, Bury St Edmunds and London, UK.

First Published 1999

ISSN 1356–1448
ISBN 1 86058 233 8

A CIP catalogue record for this book is available from the British Library.

Printed by The Cromwell Press, Trowbridge, Wiltshire, UK.

Related Titles of Interest

Title	Editor/Author	ISBN
IMechE Engineers' Data Book	C Matthews	1 86058 175 7
Handbook of Mechanical Works Inspection – A Guide to Effective Practice	C Matthews	1 86058 047 5
A Practical Guide to Engineering Failure Investigation	C Matthews	1 86058 086 6
Adapting to a Changing Railway	IMechE Seminar Publications	1 86058 189 7
Better Journey Time – Better Business	IMechE Seminar Publications	0 85298 997 0
Fault-Free Trains – A Reality?	IMechE Seminar Publications	1 86058 133 1
Developing and Growing the Business – Passenger (Railtech '98)	IMechE Seminar Publications	1 86058 182 X
Developing and Growing the Business – Freight (Railtech '98)	IMechE Seminar Publications	1 86058 183 8
Getting Trains into Service (Railtech '98)	IMechE Seminar Publications	1 86058 186 2
Infrastructure Issues (Railtech '98)	IMechE Seminar Publications	1 86058 188 9
Technology for Business Needs (Railtech '98)	IMechE Seminar Publications	1 86058 185 4
Tomorrow's World (Railtech '98)	IMechE Seminar Publications	1 86058 187 0
Train Maintenance (Railtech '98)	IMechE Seminar Publications	1 86058 184 6
New Trains for New Railways	IMechE Conference Transactions	1 86058 146 3
Railways as a System	IMechE Conference Transactions	1 86058 219 2
Train Maintenance Tomorrow... and Beyond	IMechE Conference Transactions	1 86058 095 5

For the full range of titles published by Professional Engineering Publishing contact:

Sales Department
Professional Engineering Publishing Limited
Northgate Avenue
Bury St Edmunds
Suffolk
IP32 6BW
UK

Tel: +44 (0) 1284 724384
Fax: +44 (0) 1284 718692

Contents

Towards a Fault-Free Infrastructure

C570/027/99

A systemic approach to fault-free infrastructure

D W MILLER
Halcrow Rail, London, UK

SYNOPSIS

This paper describes the approach used by the author to assure that the New Southern Railway Project in Sydney, Australia achieved the required levels of safety, reliability and availability.

It examines, by practical example, some of the pitfalls to avoid in infrastructure project definition and delivery processes that are often the root cause of faults. Of primary interest are the many interfaces (machine-machine, man-machine, man-man) which need to be identified, configured and managed.

In particular, the paper will demonstrate how fundamental infrastructure design inputs need to be derived from these interfaces and from ultimate operating requirements and not just from engineering or construction requirements.

The approach used by the author is derived from railways being socio-technical systems, a crucial aspect that is so often not recognised or under played. The paper concentrates on passenger railway systems but the philosophy is universal.

1 INTRODUCTION

At the heart of the issues described in this paper is the postulation that the cause of many faults is deeply rooted in a failure to take account of the total system within which our engineering products must operate. This includes the lack of attention paid to interfaces within this total system and the general absence of integrative mechanisms to manage these interfaces.

In particular the concept of infrastructure which is completely fault free perhaps should be challenged. Fault free infrastructure, given the vagaries of human behaviour, is likely to remain a dream for which we aim but never quite reach.

Behind much of what is discussed here is the notion that faults will occur but that action needs to be taken to prevent these affecting critical business success factors.

2 DEFINITIONS

At this point it may be useful to note that the words systematic and systemic do not have the same meaning. Systematic means being methodical. A systemic approach should be systematic; but it is much more than this.

The Oxford Dictionary definition of fault is "*defect, blemish, in character, appearance*". Perhaps a better definition is one which is closely linked to the classical definition of reliability, which is "*the ability of an item to perform a required function under stated conditions for a stated period of time*". Therefore in considering faults we are really referring to loss of function. We also know from experience that the degradation of a fault to an accident and hence to a disaster can often be rapid.

This paper concentrates on the more dramatic type of faults or loss of function; those which can lead and have led to serious accidents or disasters. Unfortunately, there are too many examples of these.

3 SOME DRAMATIC EXAMPLES OF FAULTS

3.1 Space Shuttle Challenger, January 28 1985

The space shuttle exploded less than one minute after launch. As later inquiries revealed, this was an inevitable accident that had been predicted by the designers for years. As a story of the "human factor" undermining probably the most "intelligent" organisation in the world this cannot be beaten. The subsequent Presidential Commission described the launch as an "accident rooted in history".

3.2 Channel Tunnel Fire, November 1996

The design and provision of state-of-the-art infrastructure and equipment does not in itself resolve the problem of human fallibility and complacency. The most alarming aspects of the inquiry into the November 1996 fire in the Channel Tunnel[1] relate to the Eurotunnel management. The 36 recommendations were categorised:

Fixed equipment	7
Rolling stock	4
Performance of staff, emergency services	25

Many of the procedural difficulties and deficiencies were known from previous incidents and/or had been revealed by a major safety exercise eight days prior to the incident and a safety audit some 4 months prior to the incident, but had not been addressed or corrected.

3.3 Kings Cross Underground Station

Much of the modern approach to rail safety came from the findings of the Investigation into the Kings Cross Underground Fire[2], in which 31 people died. Many of the findings related to organisational interface and management issues, for example:

- safety responsibility was divided organisationally, ie no one person in London Underground Limited (LUL) was charged with overall responsibility for safety
- there existed no managed safety programme which enabled hazards to be identified and either mitigated or eliminated
- the response of staff was uncoordinated, haphazard and untrained
- the level of co-ordination between LUL and the emergency services was almost non-existent, with particular respect to escape routes out of the station.

4 PASSENGER RAILWAY SYSTEMS

Passenger railway systems are complex in both their technical and human organisation. Underground passenger railway systems and stations, in particular, require the consideration of extensive interactions between physical systems and procedural requirements. They are living, breathing, real-time operating systems containing both passengers and operator personnel.

Where management of these interactions relies on complex organisational interfaces, there is a considerable risk that safety and reliability may not be the primary and common focus of all the organisations involved.

The responsibility of the passenger railway industry can best be summed up in the words of the Channel Tunnel Safety Authority [1]:

> *"The objective of the designers, constructors, operators and regulators of a transport system must be to ensure that it provides an acceptable level of safety for the staff and users of the system. Such a level of safety depends upon well-designed, reliable equipment operated in accordance with clear and effective procedures, particularly in the management of emergencies. But no transport system can be made completely safe, involving as it does a complex interaction of engineering, technology and people."*

Note that this does not completely rule out fault. It refers to an acceptable level of safety rather than an absolute level of safety. Most importantly it links this to the complex interactions which operate in the whole system, that is between engineering, technology and people.

In the development of new railways or extensions to existing railways it is tempting for engineers to focus on physical solutions to system design, ie the engineering and technology, without fully defining the procedural and organisational interfaces required for the overall operational integrity of the railway.

Concentrating on physical design of parts of the system generates risks of poor interfaces with the procedural requirements. This may cause unforeseen infrastructure faults and hence operational safety and reliability problems.

5 FAULTS : HOW IMPORTANT ARE SYSTEM INTERFACES AND CHANGES?

You may feel that too much emphasis is being placed on interfaces and their potential for causing infrastructure faults. Their importance is crucial.

The Inquiry into the Flixborough Chemical Works Explosion, which killed 35 people, emphasised the impact that poorly managed or poorly understood change can have, in this case with devastating consequences. The Inquiry also made clear the need to properly understand the risks in an operation and the way in which additional risk might be imported through a permanent change to the operation, and the even more risky transitional arrangements that can apply whilst change is being implemented.

Turner and Pidgeon[(3)] describe a systems view of the organisational preconditions to technological faults, accidents and disasters, in particular a seminal "man-made disasters" model. This postulates that the cause of many large-scale faults, accidents and project failures can be traced to the key interaction between technology and organisational failings. A primary organisational issue is information and communication difficulties, eg failures to assemble, disseminate and assess available warnings about hazards. These problems can be compounded if organisations are resistant to learning the full lessons from past incidents and mistakes.

This is exacerbated by difficulties which can be traced back to decisions that are taken (or avoided) at much earlier stages of the development of the infrastructure. Many major issues that bedevil infrastructure are structured into it from the start. This can be due to a strong desire to "get on with it" in the early stages, often leading to a shortfall in planning. Critical out of these issues are organisational arrangements.

6 THE NEW SOUTHERN RAILWAY PROJECT, SYDNEY, AUSTRALIA

6.1 Project Description

The New Southern Railway (NSR), design and construction of which commenced in July 1995 and will be completed by May 2000, will extend the existing Sydney suburban railway network with a new 10km twin-track mainly underground railway between Central Station and the East Hills Line, near Turella.

Four new underground stations and one surface station will be provided, including two at the Domestic and International Terminals of Sydney Airport.

The total estimated design and construction cost is £ 350 million. Design, construction and commissioning was contracted in 1995 to a Joint Venture (TBJV) of Transfield of Australia and Bouygues of France.

During a 30 year concession period from April 2000 to April 2030, the four underground stations will be leased to the private sector Airport Link Company (owned by Transfield and Bouygues) who will own, staff and maintain these stations. During the same period Transfield will maintain the track and tunnel infrastructure, although funding for the design and construction of this element is through a conventional contract with the New South Wales (NSW) Government. The NSR is therefore a quasi Build Operate Transfer scheme.

6.2 Unique Features of the New Southern Railway Project

6.2.1 Operational innovation

The NSR includes several innovative features relative to the existing system:

- private operation of stations
- ventilation control and smoke management in the tunnels
- state-of-the-art communication system
- stations with an airport "environment"
- fire detection and protection systems in tunnel
- double track, single bore tunnel

Whilst being a small extension to the existing suburban railway system, the NSR will require new working practices and a change in some of the existing operating procedures. This made it important to define and agree the roles and responsibilities of all the parties involved at both a strategic and detailed level.

6.2.2 Passenger types

With two of the NSR stations located at Sydney Airport the passengers using the suburban railway system in Sydney will comprise, for the first time, a mix of regular commuters, visiting business persons, infrequent users and tourists.

The last three types will include many differing nationalities and cultures and have significantly different expectations regarding levels of customer service compared to regular commuters.

6.2.3 Self regulation

The self-regulation regime and the 30-year concession period brought a new level of responsibility, driven by the need for a longer-term business approach rather than the short-term view normally associated with construction contractors.

6.2.4 Industry capability

As long ago as 1990, a study [(4)] of contractual, human resources, management practices and construction methods in the NSW construction industry reported:

> *"its findings demonstrated the limitations imposed on its own thinking by the industry itself - its focus is largely on contractual solutions to commercial and human problems".*

It could be suggested therefore that the NSW construction industry was perhaps less than ideally equipped to deal with the NSR and some of its unique features.

6.2.5 ***Restructuring of the NSW rail industry***

The client can rarely be regarded as a unitary concept, ie a person, but as a complex system of interest groups, some congruent, some competing. These different influences within the client can critically affect the effectiveness of the infrastructure that is being developed.

In the case of the NSR, this was exacerbated by a major restructuring of the NSW railway industry, and hence TBJV's immediate client for the NSR. The previously vertically integrated state government owned railway was restructured, in 1996, as follows:

- a passenger station and train operator
- a track infrastructure owner
- a railway maintenance and works provider
- a freight train operator

This took place after the project commenced and placed the prime interest in the project with two newly formed organisations that had competing commercial interests.

7 INTERFACES, INTERFACES, INTERFACES....

Interfaces would have existed with the previous vertically integrated rail organisation. However, the interfaces following restructuring were more commercially and contractually biased than before and were at an immature stage with respect of the change in culture that they were intended to stimulate.

The critical importance of interface management for the NSR was also recognised by the NSW Rail Safety Act 1993 and the Notice of Accreditation (NoA)[5] issued to TBJV under that Act, which specifically stated:

> *"Interface Co-ordination*
> *The TBJV shall submit an interface co-ordination plan signed by all participating parties and which clearly delineates the respective parties' responsibilities regarding the overall design, construction and ultimate operation of the NSR.*
>
> *The plan shall define, but be not limited to defining:*
>
> - *all parties involved in the design, construction and operation of all aspects of the NSR;*
> - *the respective scope of responsibility of each party;*
> - *individual interface relationships between parties, including scope of relationship;*
> - *Key events relative to each party and their respective implementation and completion dates."*

8 THE PROBLEM AND ITS CAUSAL FACTORS

8.1 The Problem

The fundamental problem identified by the author in early 1996 was that all organisations participating in the NSR had internal and external relationships and interfaces which were focused solely on the management and administration of the design and construction contracts. There was no initiative by any organisation to manage the integration of the interfaces.

This focus was the antithesis of the approach that would be required to ensure that the final product of those contracts would be the integration of an extension to the existing Sydney suburban railway system that could be operated safely and reliably.

This did not assume that the infrastructure would be fault free but that faults could be restricted to those that were neither safety nor reliability critical.

7.2 Causal Factors

7.2.1 Effect of early client decisions

The contract documents for the design, construction and commissioning phases were drafted with an almost single focus on physical infrastructure. The contract documents barely recognised the need to integrate the NSR into the existing system, the possibility of change in operational requirements during the design and construction phase and the fundamental important which interface management and procedural aspects play in achieving a safe and reliable railway.

7.2.2 Effect of client background

The client had limited experience of large-scale infrastructure project management. There was also no systematic, structured, integrated approach to rail safety, nor was there a statutory requirement to do so. For this reason and because their advisers concentrated on the administration of contracts the need to adopt a more *"system thinking"* approach only became apparent very late in the design and construction phase of the project.

7.2.3 Understanding of all factors affecting the client role

The reorganisation of the NSW rail industry brought with it a general sense of confusion regarding *"who owns what, who controls what"*.

There was a general lack of appreciation of just how much this situation differed from the previously vertically integrated railway and how this difference complicated the organisational interfaces and interactions through which a safe and reliable extension to the existing railway could be achieved.

7.2.4 Effect of existing organisational conflicts and behaviour

There was a tendency for the newly formed state owned rail organisations to *"go it alone"*, in effect ignoring the complex organisational interfaces which exist for the operation and maintenance of the NSR. Dealing with a new station operator, another major (federal) government agency, a new railway infrastructure owner and a new infrastructure maintainer required a different approach and management skills.

7.2.5 Recognition of the past, present and future of the client
Until the need for an integrative device for the operation of the NSR was identified, there was practically no recognition of the potential impact of interfaces on faults. Putting this another way, the project needed to be redefined from *"a project as a project"*, ie short term and inward looking, to *"the integration of an extension to the existing system"*, ie long term and outward looking.

This reflected the wider goals and objectives of the project. In this change of emphasis it also became necessary, for the first time, to define the:

- business product (the level of service required for safe and reliable operation)
- business case (the contribution of each organisation individually and collectively towards achieving the business product)

8 THE SOLUTION - AN INTEGRATION STRATEGY

8.1 Phasing
In the first two years of the NSR project (1995/96) the principal effort in TBJV was to mobilise and commence the major infrastructure works (tunnels and stations) , including design. This *"holes in the ground"* phase was essential for successful completion of the project. However, with all major construction fronts underway a target of mid 1997 was appropriate for a change in emphasis to *"railway systems and engineering"* and a more holistic approach to delivery of the final product.

8.2 Internal to TBJV
This involved progressive implementation of integrated, systematic, analytical methods related to railway safety, reliability and operational aspects, viz:

- System Assurance Plan
- Operations and Maintenance Objectives
- Document Development Plan
- Risk Assessment and Treatment
- Fire Safety Study
- Technical Advisory Panel
- Reliability Centred Maintenance
- Interface Management Plan

Of these the key document was the System Assurance Plan which had the following objective:

> *"To ensure that the fundamental issues of safety, reliability, availability and maintainability are taken into account during the design, construction and operation of the NSR"*

8.3 External to TBJV
Many of the outcomes of the analytical methods employed by TBJV require procedural solutions from the operators and maintainers, which could not be pre-empted by TBJV. To

be successful the integration strategy had to embrace all organisations and activities and be clearly aimed at ultimate and common objectives accepted by all the organisations.

The approach chosen to stimulate this in the inter-corporate phase was promulgated through four documents, collectively referred to as the Operational Joint Initiatives Strategy (OJIS):

- Safety Management Plan Overview
- Operational Control Specification (in particular this defined levels of service in terms of infrastructure function, eg normal, degraded and emergency)
- Commissioning Management Plan
- Reliability Centred Maintenance

The OJIS had the following mission statement:

> *"To establish and manage a structured, systematic process through which the New Southern Railway can be integrated with the existing suburban railway in Sydney in a manner which is fully understood by all parties involved in the operation and maintenance of the New Southern Railway"*

9 THE LESSONS LEARNT

Proper integration demands treating the project as a complete system from the outset, ie an assemblage of interacting and interdependent entities (eg people, things, organisations, contracts) forming a unified whole. Design inputs for infrastructure must be derived from a definition of the whole system.

Systems thinking highlights the interdependencies between project elements, and therefore the need for integration. Integration is facilitated if all project participants adopt a holistic systems view of the project. That is, having a shared view of the entire set of interrelated interdependencies between all project individuals and groups.

Management systems must be established to:

- Recognise and evaluate every conceivable potential fault of products, procedures and processes and their effects on the required function of the infrastructure
- Identify actions which could eliminate or reduce the chance of the potential fault occurring. These actions should be measured, if appropriate against predetermined functional or service levels
- Document the activities involved in these exercises

Contractual arrangements for individual phases of the project must be structured to allow for this evolution such that short term interests, eg contract administration and the defence of contractual rights, do not become an unnecessary barrier to integration.

The development of complex infrastructure must at all times remain focussed on the end objective, ie operational requirements. At no time should the focus be allowed to drift to short term interests. Such interests have a place but must always be kept in context.

The vital task is often not the engineering, the design or the construction and installation of infrastructure but the management of interfaces and interactions that have a fundamental effect of the function of the infrastructure.

10 RECOMMENDATIONS

The design, construct and commission phases must include continual validation, testing and review of operational requirements.

Integrative mechanisms are essential between those organisations that will design and construct the facility and those who will operate and maintain it. These integrative devices should recognise the differing levels of operational (or project) knowledge at various stages and be aimed at ensuring a gradual transition at the appropriate time.

As Reyes et al[(6)] noted:

> *"Systemic means looking upon things as a system or as a dynamic whole. Systematic means being methodical. Systemic means seeing a pattern and inter-relationships within a complex whole; to see events as products of the working of a system. A systemic approach should be systematic; but it is much more than this."*

And, as Ackhoff [(7)] emphasised:

> *"Performance of the whole is not the addition of the performance of the parts, but it is a consequence of the relationship between the performance of the parts. It is how performance relates, not how it occurs independently of the other parts. That is what systems (systemic) thinking is about."*

11 CLOSING REMARKS

I hope that this paper has given you food for thought and presented a different perspective on the achievement of fault free infrastructure. To close I leave you with this quotation from John D Rockefeller, Snr.

> *"....if you want to succeed you should strike out on new paths rather than travel the worn paths of accepted success...."*

and a slightly modified version of an extract from an article by John Evans, an engineer and psychologist[(8)]:

> *"Faults tend to be characterised by their consequences rather than their causes. But if reoccurrence of faults, or better still prevention, is to be achieved then it is causal factors that need to be addressed. Incorporating good practice into tasks as a matter of routine will reduce the risk of faults and create a working environment better able to tolerate the variability, vulnerability and occasional unpredictability of normal human behaviour. Treat human error like any other problem. Find out why it is happening and fix it."*

13 REFERENCES

(1) CHANNEL TUNNEL SAFETY AUTHORITY (1997), *Inquiry into the Fire on Heavy Goods Vehicle Shuttle 7539 on 18th November 1996.*

(2) THE DEPARTMENT OF TRANSPORT (1988), *Investigation into the King's Cross Underground Fire, Her Majesty's Stationery Office, UK.*

(3) TURNER, B A and PIDGEON, N F (1997), *Man-made Disasters,* Butterworth – Heinmann, 2nd Edition.

(4) NSW DEPARTMENT OF PUBLIC WORKS AND SERVICES (1996), *NSW Government Green Paper. The Construction Industry in New South Wales: Opportunities and Challenges.*

(5) NEW SOUTH WALES DEPARTMENT OF TRANSPORT (1966), *Notice of Accreditation, Transfield Bouygues Joint Venture.*

(6) REYES, J, BEARD, A N and CLARKE P J (1997), *A Systematic Approach to Fire Safety Offshore,* Proceedings of an International Conference on Fire at Sea, Royal Institution of Naval Architects, London, November 1997.

(7) ACKOFF, R L, *The Second Industrial Revolution,* Manuscript, Wharton School of Finance and Commerce, University of Pennsylvania.

(8) EVANS, J, *Human Error Is Not An Explanation,* Quality World, 1998.

It's All About Performance and Its Role in the Efficient Infrastructure

C570/018/99

The importance of condition monitoring in providing fault-free infrastructure

K FRY
AEA Technology Rail, Derby, UK

SYNOPSIS

Many approaches to railway privatisation around the world have seen the commercial separation of the vehicle from the infrastructure. Whilst this forms a convenient administrative boundary there is no escaping the engineering interaction between the two. The poor condition of one element can directly cause significant exceptional costs to be incurred by the other. In circumstances where the condition of another company's assets can have such a significant affect on your own costs how can this be effectively managed? One answer lies in the use of equipment designed specifically for asset condition monitoring and protection.

AEA Technology Rail has been involved in condition monitoring in the UK railway industry for over twenty years, producing systems that cover both infrastructure and vehicles. Since UK railway privatisation the importance of managing the vehicle-infrastructure contractual interface has increased significantly. This has resulted in a huge demand for suitable monitoring systems, often requiring the development of new technology.

This paper describes some of the systems presently being used in parts of the UK railway and overseas that offer significant potential for improving infrastructure fault prognosis and prevention. In particular systems that cover the wheel-rail and pantograph-overhead line interfaces are described.

1. INTRODUCTION

The application of condition monitoring offers significant potential for improving many aspects of railway operation and maintenance. Since railway privatisation the importance of condition monitoring has increased significantly, in particular systems for assessing the track-train interface, a key contractual boundary between parties, and in reducing penalty payments for service delays. This paper will describe monitoring systems aimed at improving asset performance by identifying developing faults before they affect the train service and reducing cost by allowing more focused maintenance.

2. PREVENTING DE-WIREMENT

The costs of service disruption and repair following de-wirement can be huge and after the event it is not always easy to identify whether the fault lay with the pantograph, the infrastructure, the local weather conditions or a combination of all three.

Pantographs need regular maintenance to ensure that the carbon contact strips are in good condition, that aerofoils are set at the correct angle and that the uplift force is within prescribed limits.

Static measurements do not guarantee the performance of the pantograph when in operation at speed. But experience has shown that the overall condition of a pantograph can be determined by measuring the movement it induces in the overhead wires when passing a measuring location at speed. Panchex® equipment located at the trackside and on the overhead line equipment (OHLE) is designed to do this.

The trackside Panchex® equipment comprises an uplift measuring potentiometer on the OHLE and an isolating fibre-optic telemetry link to bring the potentiometer signal down to a line-side cabinet. Track switches are used to detect the presence of a train and additional equipment is used to measure wind speed, direction and ambient temperature. Automatic Vehicle Identification (AVI) equipment is used to uniquely identify the train.

Following processing, the lineside equipment transmits information to a display system at nominated locations via a modem. The trackside equipment may be placed at several strategic locations in a railway network and information display systems' in several control rooms in a region.

If a pantograph produces an uplift that is outside prescribed limits, the system sounds an alarm in the local control room, allowing a controller to slow or even stop the train before any damage is caused. In addition, both the controller and vehicle maintainer may retrieve historical information on the condition of particular pantographs, allowing predictive maintenance such that alarms become infrequent events.

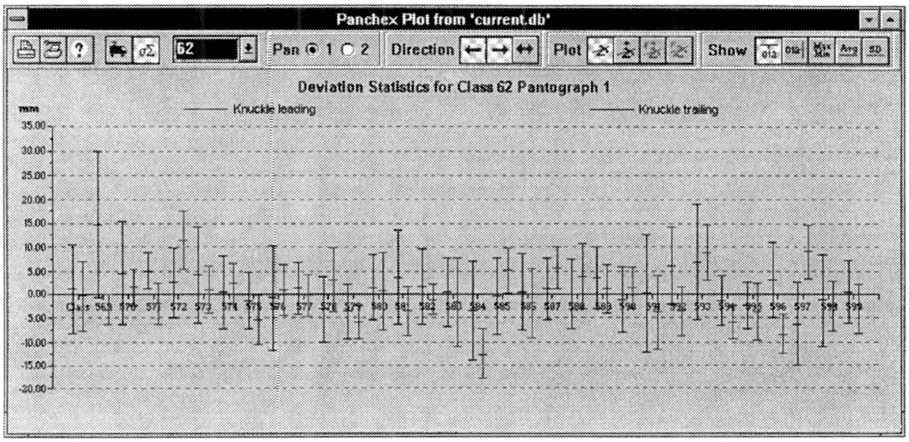

Panchex® is presently being implemented nationally in the UK with 25 trackside systems and over 30 control room and maintenance depot displays. Before long every train operator will have access to this valuable information.

Faulty pantographs are only half the story, and means to detect faulty infrastructure are equally important and this has led to the development of OLIVE®. This is a monitoring system fitted to an (otherwise standard) pantograph equipped vehicle, running in revenue earning service.

It can detect two types of fault; excessive acceleration (an impact) on the pantograph head; and excessive movement of the OHL off the centre line of the pantograph (stagger exceedence).

Impacts are measured by a fibre optic accelerometer mounted on the apex frame of the pantograph. Stagger exceedences are measured by a mechanical switch on each horn guide of the pantograph, which operates a fibre-optic interrupt. The locations of any faults are identified by an onboard GPS system.

Fault data are saved in the on-board computer system and downloaded regularly by mobile telephone to the same information displays already in place for Panchex®. The information consists of the location and severity of all impact and stagger exceedence events. Events from the databases are plotted on a map and can be grouped to identify common faults and ranked in terms of most recent, most severe etc.

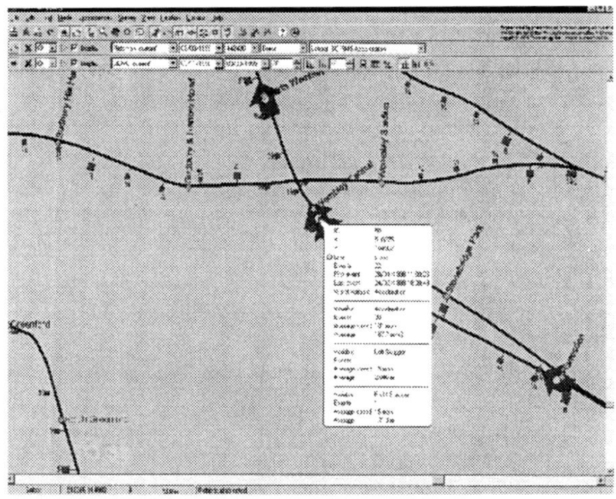

Two systems are presently fitted to service vehicles owned by Virgin West Coast and GNER train operating companies operating on the UK's East and West coast main lines. A number of additional systems are presently being installed.

Information from these OHLE condition monitoring systems has already demonstrated its potential and as their implementation continues there is every likelihood of a significant reduction in the number of future de-wirements.

3. RAIL DEFECT/BREAK REDUCTION

The management of rail defects and the prevention of breaks to avoid service disruption is a major issue at the moment. Rail flaw detection (RFD), is presently done by manual methods and is very slow and expensive. Increasing emphasis on worker safety is requiring less red zone working and greater volumes of traffic are reducing the potential for possessions, so the opportunity for manual inspection is reducing. With the number of rail breaks on the increase experience shows that present methods are unable to prevent a significant number of faults with consequent affects on safety and performance. There is an obvious need for improved methods of working.

There are two areas in particular where monitoring can help. Firstly by prevention, reducing the cause of the problem at source. Secondly by more productive, automated rail flaw detection and recording. These two aspects are covered in the sections below.

3.1 Prevention

Rail defects result directly from excessive wheel loads caused by flats and out of round wheels. The identification and removal of these from the network will significantly help to reduce the problem at source. AEA Technology Rail's WheelChex™ system is one such approach, designed to help identify faulty wheels and protect the infrastructure against the damage they cause.

The system uses strain gauges attached to the rail to measure the vertical wheel loads of passing trains at full line speed. Typically eight sleeper bays are instrumented, with the sleepers specially arranged to provide the best coverage of successive wheel revolutions. For each wheel both the mean and the peak load is determined.

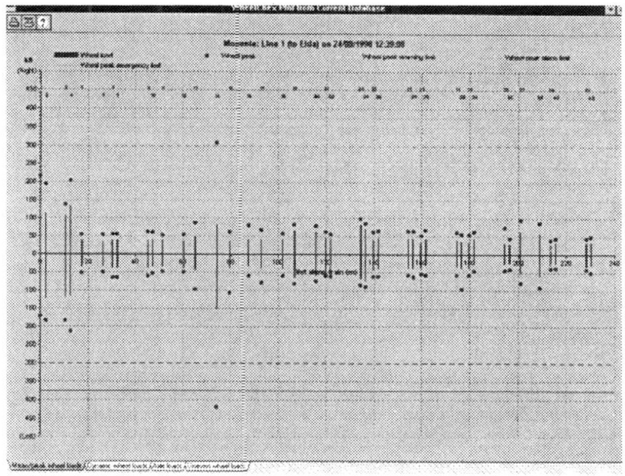

The measurements are processed and passed by telephone link to a display computer at a control room – potentially the same computer as the one described above for de-wirement detection. If a high wheel load is detected, an alarm is sounded and a message is displayed advising on a relevant course of action. A display of the train's peak and quasi-static loads is recorded and can be displayed graphically.

In addition to protecting the infrastructure against faulty wheels the information provided can be used in the assessment of total traffic tonnage and track component life prediction.

3.2 Defect detection and management

Some rail defects are inevitable however and moving to a more automated and intelligent rail vehicle based means of detection opens up the possibility for significant improvements in the way flaws are managed. The UK railways actually seem some way behind the rest of the world in this respect. Elsewhere a range of vehicle borne options are used with varying levels of technological sophistication.

Of critical importance is for the system to allow differentiation of action based on an indication of defect size. Without this, staff are faced with unmanageable requests for remedial action and the only outcome of running faster is in finding more defects, not all of which need attention.

The right automated approach will improve the control of rail flaws in four important directions:

Quality: the ability to detect, size and locate a flaw will be improved significantly.

Safety: a significant reduction in unsafe manual track working.

Productivity: systems can now operate continuously at speeds in excess of 30 km/h.

Auditability: each run can be recorded in detail allowing it to be replayed in the office.

In recent months AEA Technology Rail has acquired the rights to a RFD system developed in Australia and successfully sold to train operators in the Far East including Japan and Hong Kong.

Where RFD equipment of this type has been used the results show significant improvements have been obtained, with typically a twenty-fold increase in productivity compared to manual means.

Speed of measurement is only one benefit however. As the system records all of the results it is possible to check again later in the office, guaranteeing that nothing is missed. In this way the results are auditable and can be demonstrated to a third party. Detection is accurate enough for defects to be sized sufficiently to allow prioritisation of remedial action. Many defects can be seen to be small enough to be safely left. Historical runs can be compared with new on the same section of track to monitor flaw growth.

4. TRACK QUALITY MANAGEMENT

4.1 Ride measurement

The measurement of track geometry using High Speed Track Recording Coaches (HSTRC), whilst accurate is usually too infrequent to keep on top of developing faults and allow sensible workload planning by contractors. Track faults causing high vehicle body accelerations can develop between HSTRC runs with rough riding reported by drivers. Train delays may then result when temporary speed restrictions are imposed and incidents investigated and the fault repaired. To help manage this situation many parties are now performing supplementary manual ride measurement runs. To support this approach and improve the efficiency of the process AEA Technology Rail has developed RideMon™, a completely automated, qualitative track quality measurement system.

RideMon™ consists of an office-based display linked by cellular radio to one or more permanently mounted on-board ride monitoring systems fitted to service vehicles. The on-board computer records the magnitude and location of high lateral and vertical accelerations and also provides measures of ride quality over 1/8 mile (200m) sections of track. The system is completely automatic from train to desk.

The resulting information provides an easy to use display of poor ride quality sections, presented via a colour-coded map. Historical comparison of particular track sections can be made allowing developing faults and the effectiveness of remedial action to be monitored.

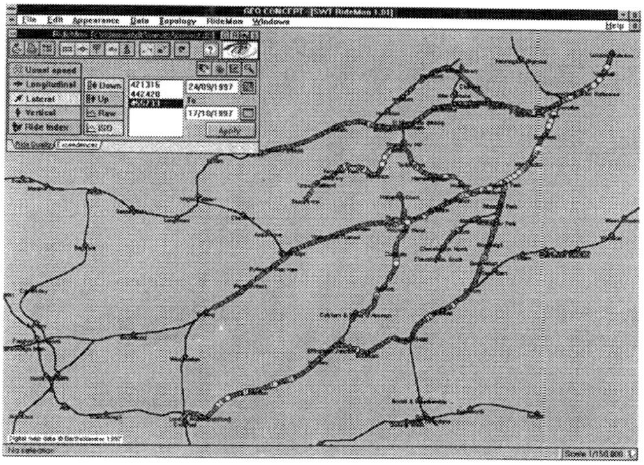

The use of ride quality monitoring on service trains is a powerful and low cost method of assessing the general condition of the track over which a train operates.

The information is of obvious benefit to the infrastructure owner and maintainer but is also useful to the train operator and maintainer. Results from the vehicles are disseminated to all parties automatically whilst respecting all commercial confidentiality.

RideMon™ is presently fitted to three multiple units operated by South West Trains Limited and two High Speed Trains operated by First Great Western. Further trains for Thames, Connex and Silverlink will be fitted shortly.

5. LOW ADHESION IMPROVEMENT

Train operators rely on high acceleration and deceleration rates to operate a satisfactory service. The poor braking and traction performance of trains under low adhesion conditions causes significant service delays and potentially affects safe operation. Low adhesion can occur at any time in the year but is a particular problem in Autumn.

To help manage this problem the cross-industry Adhesion Working Group and AEA Technology Rail has developed a Low Adhesion Warning System (LAWS™), which provides a real-time measure of low adhesion events and locations. The system comprises an on-board computer that monitors train-wide WSP (wheel slip/slide protection) activity during power and braking. Any WSP activity is recorded and classified against, speed, duration, distance and brake step or power notch in order to identify severity. The determination of geographical position is achieved using a combination of satellite navigation and distance derived from the train speed input.

The information from each vehicle is automatically transmitted in real-time to an office based Low Adhesion Mapping computer. This builds up a comprehensive geographic picture of low adhesion conditions.

The information is then used to warn drivers of low adhesion sites before starting their shift and to better target remedial action to the areas and times shown to be most in need.

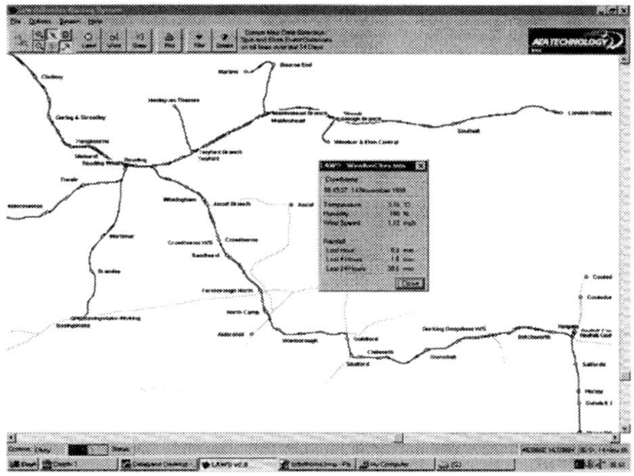

Recently work has been undertaken to link data from remote weather monitoring stations to the information display. The aim being to provide advance warning of the development of the weather conditions known to cause the most severe low adhesion situation.

LAWS™ is fitted to four multiple unit operated by Thames Trains Limited. A number of further systems are presently being fitted to trains operated by Silverlink, Connex and First Great Eastern in time for this Autumn. As the system shares many core components with RideMon and Olive it can easily be added to the same vehicles.

6. POINTS FAILURE PREDICTION

Points are always high on the list of infrastructure assets for which failure results in a large loss of performance. Their design and the typical modes of failure make point machines an ideal candidate for performance improvement through effective monitoring and fault prediction. Accordingly, in 1994 work funded by the British Railways Board and subsequently Railtrack, started on a system for fault prognosis of clamplock points.

In order to develop an effective prognostic system, a good understanding of the different modes and mechanism of failure is required. An analysis of FRAME data and Reliability Centred Maintenance work showed that failures could be broken down into five major areas:

- High friction due to poor lubrication.
- Equipment out of adjustment.
- Failure of the detection equipment/wiring.
- Obstructions to the mechanism.
- Broken/Fractured equipment

The next stage of development was to gather real information about how measurable parameters varied under different fault conditions. About 130 individual tests, covering all the fault modes expected, were carried out using specially instrumented points at the S&T training school at Stafford. In all, data for over 900 switch operations were collected. These data showed a number of particularly interesting results:

- Switch operation time is almost independent of a developing fault.
- Pressure characteristics are repeatable for a particular switch condition.
- Pressure characteristics change significantly for different fault conditions.
- Pressure characteristics give good early warning for increasing friction faults.
- The pressure characteristic during locking and unlocking shows lock arm and detection cam movement which could be used to aid diagnosis of problems.

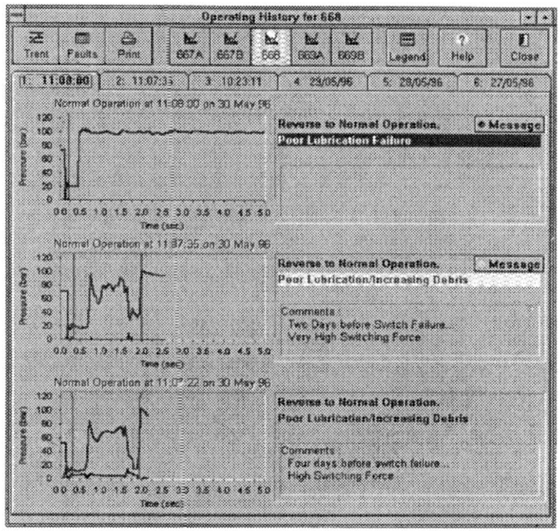

The tests showed that good prognosis was possible simply by measuring the pressure characteristic in normal and reverse operation, requiring only two pressure transducers and some digital indications of detection and pump activity. Using these results, automated analysis software was developed to identify common developing fault conditions with the expectation that early warning of failure is possible for 50% of faults.

In order to assess the performance of the system in service a test installation at Trent East Junction was established with five switches monitored. A track side computer was installed that carried out all of the data analysis every time a switch operated.

Information about developing faults and switch failures is automatically transmitted to display machines at a number of locations. The results for each switch operation can be viewed, and a list of the current faults and warnings is maintained so that the operator can monitor the situation and record actions taken. The software runs on a PC in the Windows environment and is designed to be easy to use and give a quick visual display using colour and graphics so that the operator can see at a glance if there are problems.

7. CONCLUSION

The challenge for today's railways is to improve performance whilst at the same time cut costs. It is not an easy one. With increasing traffic the consequences of faulty infrastructure and the knock on delay costs will most likely rise. At the same time the opportunities for maintenance will actually decrease whilst the need for it may rise. If we are to have any hope of meeting the fault free infrastructure challenge we will have to work smarter. In an already well optimised industry new technology and methods of working must be the key. This paper has attempted to show the important role condition monitoring systems in particular play in working towards fault free infrastructure.

C570/014/99

Self-monitoring and reporting telephones as part of a fault-free railway infrastructure

P M COOPER
GAI-Tronics Limited, Burton-upon-Trent, UK

SYNOPSIS

This paper describes a range of "SMART" telephones, developed for the rail industry, capable of providing regular health status and call data. Its main focus is a new generation customer Information Point that integrates with public address and visual passenger information systems.

A Telephone Management System (TMS) has also been developed that can monitor telephones connected to a public switched network, a private network or cellular-based fixed location telephones. Enhancements to the TMS maintenance system allow it to send operational messages to Information Point telephones to show current train information on a display panel, which is updated regularly using "silent" calls from a central point.

1. SELF-MONITORING AND REPORTING TECHNOLOGY (SMART)

By incorporating a low power microprocessor and real-time clock, it is possible for a telephone to monitor its own health status, record call details, report-in regularly and call automatically if external sensors are triggered. Careful circuit design allows such telephones to operate from the micro-amp currents that are available from the telephone line when they are on-hook. Previously, only telephones with auxiliary power, from an electrical supply, could be monitored in this way.

SMART telephones offer real benefits to rail companies through:

- Availability of telephone health status records for incident enquiries
- Immediate warnings of vandalism
- Reduced preventive maintenance costs
- Provision of exception reports to identify maintenance activity
- Availability of statistics on calls and call attempts
- Reconfiguration from a remote location.

There are some 10 000 signal-post telephones on the UK rail network. Prior to the introduction of SMART technology, their operation could only be confirmed by maintenance personnel visiting units periodically. This is clearly an expensive activity, which only guarantees the telephone is working on the day of teat. Such telephones have a typical calculated meant time between failures (MTBF) in the range 50 000 to 100 000 hours, so with vandalism and ageing effects many hundreds have to be located and replaced every year. The Telephone Management System is able to generate reports from routine polling activity to highlight only those telephones that need maintenance activity, eliminating the need for routine manual checking of every unit.

SMART telephones have been available for over a year in a range of hands-free and handset versions. Through the use of microelectronics, SMART technology can be incorporated into standard sized telephones and an example is illustrated below.

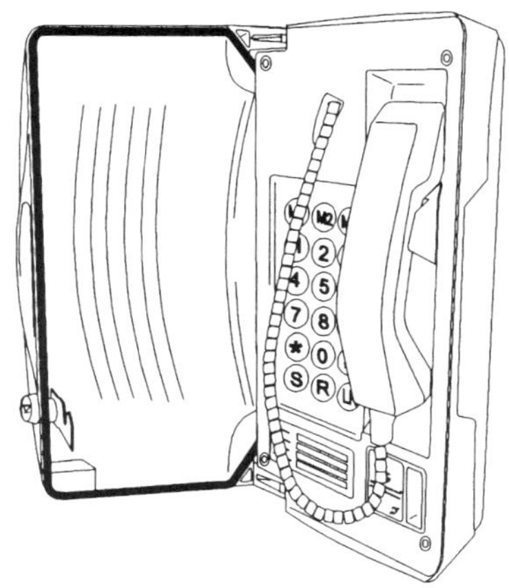

Figure 1: SMART metal-bodied weather-resistant telephone

2. SMART HARDWARE IMPLEMENTATION

The SMART phone can be configured to call in automatically, report its status and provide details of calls made. This is instigated by the real-time clock triggering a microcomputer interrupt to wake the phone from its idle state. A Telephone Management System can be installed at any location having access to the phones on a public or private network. The system of data transmission employed allows telephones on a public network to be monitored from anywhere in the world. GAI-Tronics often monitors SMART phones in Europe, Asia and the USA.

The block diagram of a SMART telephone is shown in figure 2 below.

When the SMART telephone is on-hook, leakage current is fed from the line interface through the electronic hook switch to the microcomputer and real-time clock to keep the phone "alive."

This leakage current is so low that a standard REN 1 approval is achieved. Lifting the handset causes the phone to function normally, but dialling is implemented via the microcomputer and call details are stored in non-volatile memory.

Incoming calls are answered in two stages. The first stage leaves the ringer silent, but the line has been looped and the microcomputer listens for a "Maintenance Signal." If not present, the ringer starts to sound after a delay (stage two), until the handset is lifted or the call is terminated. A remote operator can send a maintenance signal during stage one, which inhibits the ringer and allows the operator to gather data from the unit and/or change its configuration. The most common configuration changes are to re-program auto-dial button memories.

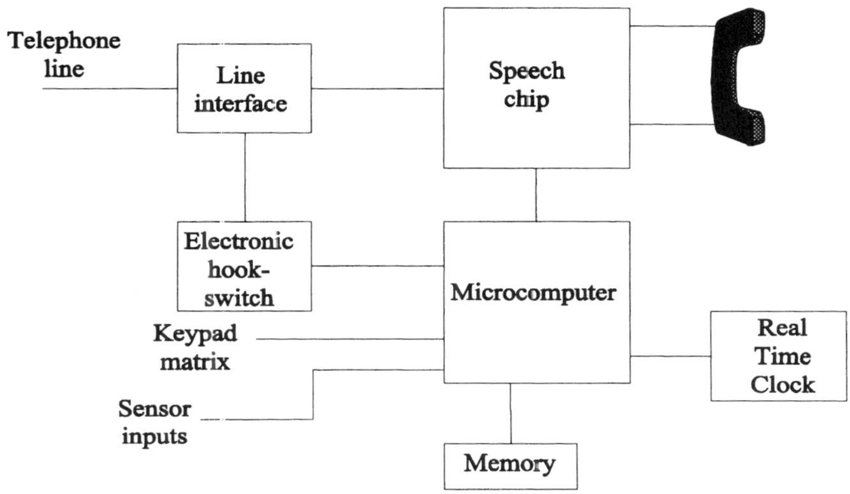

Figure 2: Typical SMART telephone block diagram

SMART telephones are not inherently any more reliable than any other type, however, these products are third generation designs, benefiting from significant railway experience with previous weather-resistant telephones. They have many mechanical features that contribute to a high level of reliability in a railway environment.

3. TELEPHONE MANAGEMENT SYSTEM

The Telephone Management System (TMS) comprises a Personal Computer (PC) running dedicated software and connected to a modem. It can be set up to monitor several hundred telephones and GAI-Tronics has one customer with over six hundred phones on a monitoring system.

The TMS allows individual telephones to be configured and the data is also stored in a PC database. Phones can be configured to call in regularly and download status and call data. Alternatively, the TMS can be configured to poll phones and this is usually more appropriate for large systems. A system operator can interrogate the TMS database and provide reports on any phone, a group, or all units. Additionally, exception reports can be generated to highlight any maintenance activity required. Reports are automatically generated daily and these are stored as text files on the computer. The daily reports can provide essential data should it be necessary for an incident enquiry, confirming phones are healthy and listing times and dates of calls made. The files can be backed-up and printed using standard PC techniques.

Exception reports can be generated to identify telephones requiring maintenance activity. Typically, the data available might include:

- Door not closed
- Handset off-hook.
- Handset not present
- Phone did not call in
- Tilt switch, or other vandal-activated sensor, operated.

The availability of good maintenance data is the key benefit for a fault free railway infrastructure. This can eliminate the need to walk the track regularly to confirm signal post telephones are working, leading to operational benefits. It also pinpoints units requiring attention, leading to faster repair and improved availability. Note that door open and handset off-hook do not leave the telephone out of service. Because the unit has an electronic hook-switch it can be returned on-hook even though the handset may have been left dangling.

Once data is stored in the TMS it can be made available to other computer systems via a network or modem connection. In this way, a train operating company or track provider could centralise the maintenance monitoring with other completely independent systems to take the maximum benefit from automatic maintenance monitoring.

4. RAILWAY INFORMATION POINT TELEPHONES

The latest development of SMART (technology) is provision of passenger information for stations and public access areas. Such a unit can be built with limited optional features into standard weather-resistant telephone cases as illustrated in figure 3 below:

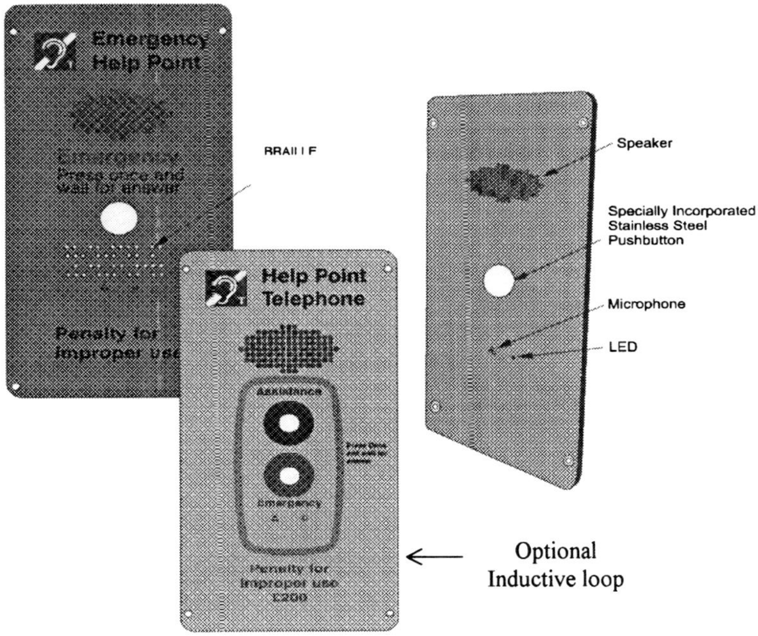

Figure 3: Hands-Free SMART Help Point Telephones

The full product is called "Information Point" and GAI-Tronics Ltd, working in conjunction with Racal Telecommunications Ltd, has developed this specifically for rail industry application. By the addition of a power supply, a second microcomputer and a display, it is possible to provide a range of passenger information assistance facilities through a single unit. For small stations such a unit can also provide remotely operated public address (PA), visual display, help-point assistance and emergency communications with remote control from a central point. With the optional facilities to provide remote PA and running train timetable data, Information Point is a cost-effective solution for passenger information at small stations in a rail network. With its internal display and speaker it can provide passenger information locally in large stations, for instance relating to one or two platforms operated by one Train Operating Company (TOC). The column is physically robust, treated with an anti-graffiti coating and can be coloured in a livery to suit either a TOC or a station specific colour scheme. Information Point has been developed as a modular family of products and the design concept is illustrated in figure 4 below:

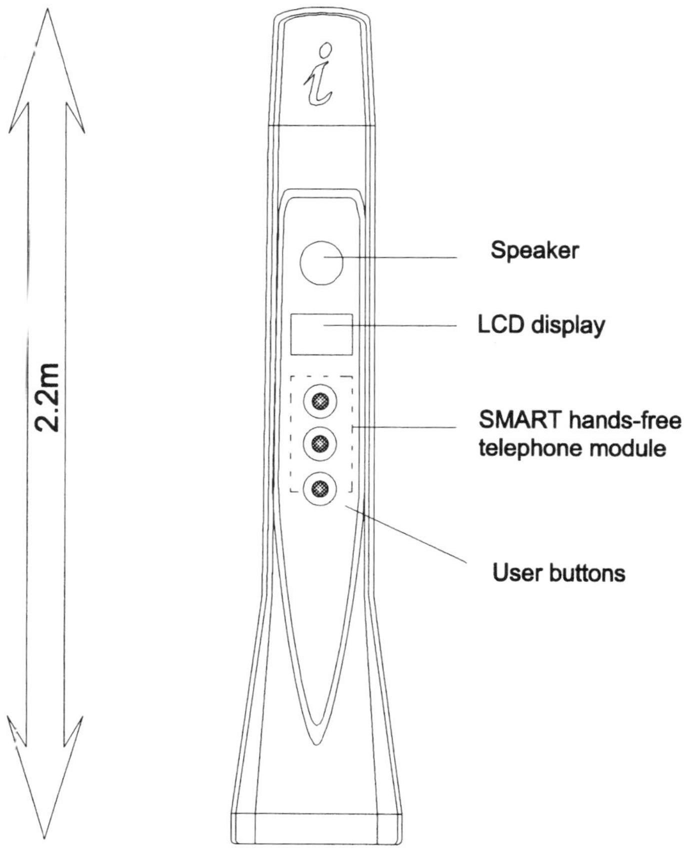

Figure 4: Information Point

The Information Point comprises a SMART hands-free telephone and a number of optional audio/visual interfaces, including LCD display, PA function, digitised speech storage and data links. The telephone module is hidden from the user and replaced with front-panel mounted illuminated buttons, microphone and speaker. The unit may be linked to a variety of other equipment, including emergency operator systems, station controllers and remote management systems. It can accept signals from a telephone line, an RS 485 data bus, and/or a station PA and can provide outputs to local display and speaker, external platform displays and an external PA system. In addition, state sensor inputs can be accepted and output contacts are available to trigger, for example, lighting and security cameras. The LCD can display up to four lines of text, which can be scrolled for long messages.

A number of user buttons are available on the Information Point which, typically, offer:

- Playback of digitally recorded audio messages
- Connection to a pre-recorded information line
- Emergency telephone connection.

A block diagram of the Information Point is shown below in figure 5:

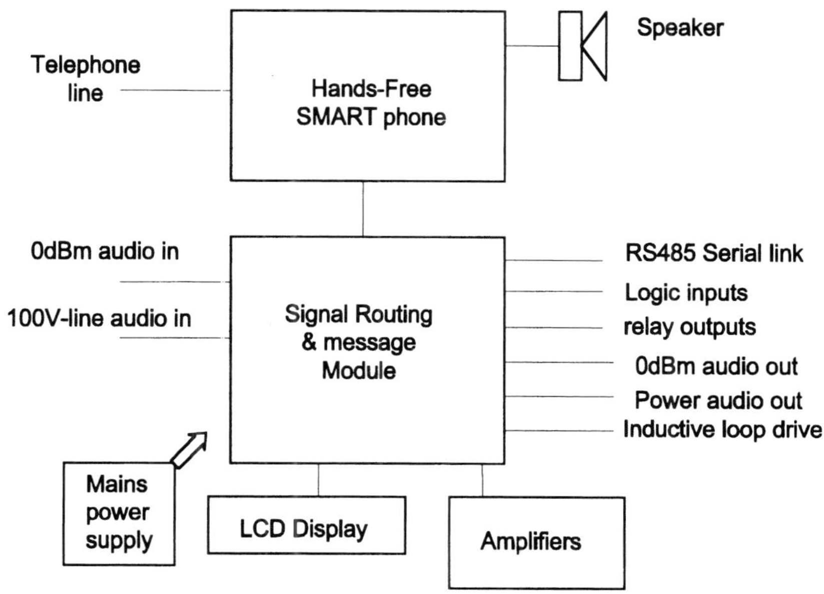

Figure 5: Information Point block diagram

The telephone module functions in much the same way as a standard SMART phone, but there is a serial link to a second microcomputer in the signal routing and message module. Depending on the configuration (set by customer requirement) audio signals can be routed from any of the audio inputs, including the telephone line, to the internal speaker, to a digitising memory and to audio outputs. Audio messages can also be triggered from the memory to play through any of the other outputs. Display information can be received either from the telephone line input or from the RS485 serial data bus input. These signals are stored locally and can be displayed on the internal LCD panel and on an external display, if required. The internal display can be split under software control to show both internally generated and externally provided messages at the same time.

5. TELEPHONE MANAGEMENT SYSTEM ENHANCEMENTS

Because the Information Point is based on SMART (technology) all the system management benefits detailed in 3 above are available. The Information Point LCD can be used to display messages stored within the unit about telephone call progress, standard "Welcome" messages or advertising material. However, because of TMS it is possible to silently download messages for presentation on the LCD and this allows the display of updated timetable information, including any delays to trains or cancellations. Providing the TMS PC has access to rail network train running data it can issue display messages to the Information Point in exactly the same way it polls telephones for status and call data.

6. AN INTEGRATED CUSTOMER INFORMATION SYSTEM FOR SMALL STATIONS

The Information Point telephone has the ability to play either internally stored digitised speech messages or to play dialled up pre-recorded messages through its loudspeaker. For an integrated approach to passenger information there needs to be a relationship between the messages that are presented on the display and any audio messages it provides. This can be achieved by adopting a similar approach to the telephone management system computer.

To ensure the Information Point always has up to date audio messages available, the "Information" button is used to auto-dial an Audio Message Forwarder computer which holds all current audio messages (as data files) for every Information Point in a group. When a passenger requests an information message, the appropriate file is routed to the Information Point as an audio signal and plays through Information Point loudspeaker. If required these messages can also be routed into the platform PA.

For the passenger requesting information, the audio message is much like dialling the "Talking Clock." The Audio Message Forwarder knows which file to route to the Information Point from the Caller Line Identification (CLI) data which is incorporated in the ring sequence by the telephone network. Thus audio is instantly presented as soon as the Audio Message Forwarder answers the call.

For large systems both the Telephone Management System and the Audio Message Forwarder computers can communicate with several Information Points simultaneously by using multi-channel modem cards.

A block diagram for an integrated customer information system is presented below:

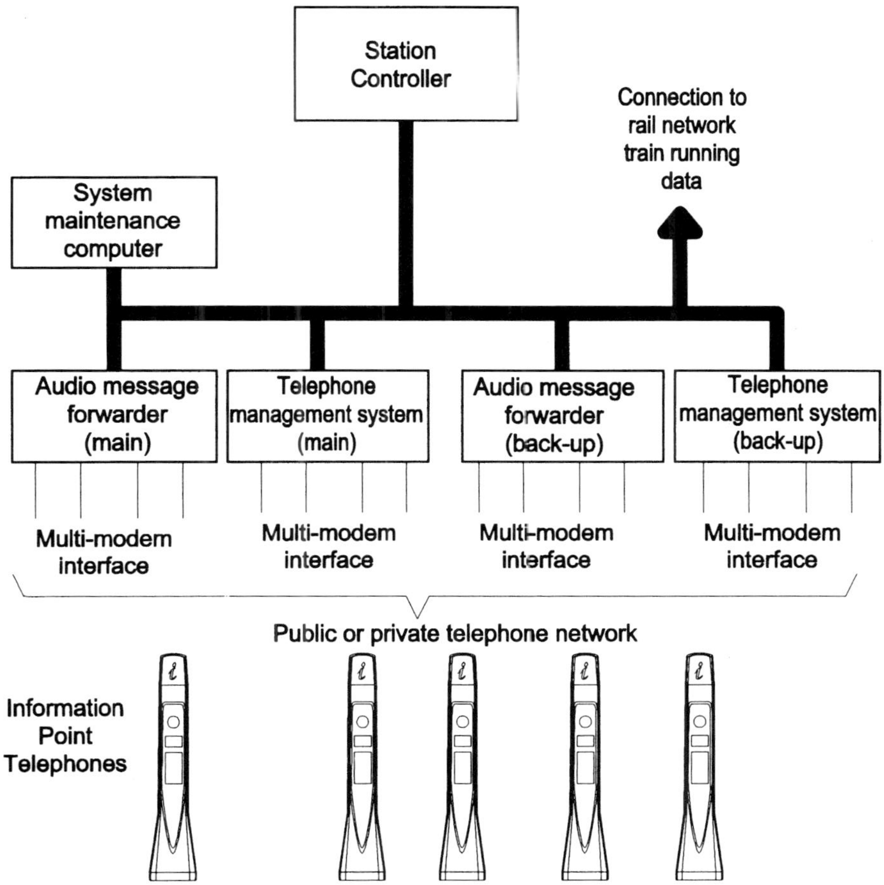

Figure 6: Integrated customer information system incorporating Information Point telephones

Rail network train running data is presented to the integrated customer information system via a master station controller computer, which prepares both audio and visual display messages from data stored locally. These messages are passed to the Telephone Management System and Audio Message Forwarder PCs, via an Ethernet link, for onward transmission to the Information Points. In the case of visual display messages these are routed to the appropriate Information Point immediately and presented on the display until updated, or until their validity expires. In the case of audio messages these are stored by the Message Forwarder for playback on request by a passenger at a specific Information Point. By adopting this approach telephone line time is minimised and the need to record audio messages at the Information Point is avoided.

The Information Point has a single telephone line that is used for visual message data transmission, daily status monitoring, customer information audio messages and emergency telephone calls. The SMART phone micro-computer in the Information Point is able to prioritise its functions so that if an emergency call button is operated any other line communication is terminated and the emergency call is dialled. In this mode the passenger has normal two-way telephone connection to the emergency operator, with the Information Point operating as a hands-free telephone.

As with standard SMART telephones, the Information Point is able to provide status information to the Telephone Management System computer through maintenance calls. For a truly integrated system the Audio Message Forwarder and the Telephone Management System feed status data via an Ethernet network to a maintenance computer which is able to report on the complete system.

To maintain a fault free infrastructure both the Audio Message Forwarder and the Telephone Management System are duplicated. The Master Station Controller constantly updates both systems. Should either audio or display message computer fail the telephone calls can be diverted to the back-up computer under the control of the Maintenance Computer.

7. SUMMARY

Through the introduction of low power microcomputers into telephones a range of new facilities can be added, offering benefits to both passengers and the operators of the railway infrastructure. The ability to gather status information from all the phones on a network reduces maintenance costs and increases telephone availability. Note that the introduction of SMART technology does not increase the reliability of telephones in itself, it is pin-pointing faulty units that allows a fault-free infrastructure to be achieved. The ability to integrate displays and PA-type facilities with a SMART phone provides passengers with better travel information, particularly at small stations on a rail network. An integrated station controller network allows up to the minute information to be available to station Information Points, ensures system availability through the duplication of key system components, and minimises service activity through the use of a maintenance computer.

Paul M Cooper.

The author is a Chartered Engineer and Fellow of the Institution of Electrical Engineers. He is Engineering Director of GAI-Tronics Ltd., where he has worked for ten years on the development of telephones and communications systems for the rail industry. He graduated from North Staffordshire Polytechnic, now Staffordshire University, in 1975.

C570/016/99

A new high-performance switch actuation system for a high reliability railway

S SPURLOCK
Industrial Actuation Division, Claverham Limited, Bristol, UK

SYNOPSIS

This paper describes how a draft specification for the design of S&C has been introduced to the UK railway industry, and how a new high reliability switch actuation system is being developed for tomorrow's high reliability railway. It focuses on the innovative High Performance Switch Actuation (HPSA) point machine that has been developed as a sub-system of the overall High Performance Switch Actuation Sub-System (HPSS).

The project has used a systematic approach to requirements capture, design, development and testing of this innovative piece of railway equipment. Results to date from the prototype trial prove that the HPSS will make a major contribution to the fault free railway of the future.

1 BACKGROUND TO PROJECT

Reliability of the British railway infrastructure has been at the heart of the debate over the future of the network for many years. In 1991, British Rail's statistics highlighted the fact that on average each of the 18,000 powered points in the railway network failed once every 9 months. Even now, eight years later, a recent report commissioned by the Office of the Rail Regulator (1) shows that the ubiquitous "points failure" causes nearly 25% of delays in the track and structures category.

Statistics show that this situation is improving, however the improvement may not be sustainable as it has been achieved with higher levels of inspection, maintenance and repair of Point Machines whose vintage goes back to the 1950s (GEC and Westinghouse machines), or the 1970s (BR designed Clamplock).

InterCity and Network South-East initiated a project in 1992 to produce a set of performance specifications for the design and development of Switch and Crossing systems (S&C).

This initiative was a major innovation for the railway industry that had traditionally relied upon the Permanent Way engineering staff to install the track, switches, and crossings, and the Signalling and Telecommunications (S&T) staff to install signalling equipment including Point Machines.

The specification project team comprised a number of railway industry experts, including representatives from British Rail (including BR Research), and the railway industry supply base (including GEC, Westinghouse, Balfour Beatty, and ABB). The Project Manager for this project was Fairey Hydraulics Limited (now Industrial Actuation Division, Claverham Limited), who with their aerospace and defence industry background had experience of systems engineering, multi-disciplinary Project Management, and specification writing.

This project, which lasted nearly 18 months, demonstrated that unreliability of S&C systems was caused by a lack of "system" design as opposed to the inherent unreliability of any single component within the system. The emphasis therefore of the new performance specifications was to ensure that the design and development was considered as a complete system, with defined boundaries, interfaces and performance requirements.

The S&C specifications, known as the ES3200 series, were adopted as draft standards in May 1994 following endorsement by Railtrack's Safety Assessment Panel (SAP). The SAP committee recommended that the specifications could be validated by means of the design and development of a trial S&C system.

Following their involvement with the ES3200 project Fairey Hydraulics Limited identified the opportunity to introduce a step change in the way that S&C systems were designed and developed. A joint project team was set up with Balfour Beatty Rail Engineering Limited, to design a high reliability S&C system, based around a typical UIC rail section, and including a new type of Point Machine. The project's intention was that the S&C system would be designed and developed in compliance with the requirements of the ES3200 specifications.

This paper focuses on the design and development of the HPSA unit as it represents a fundamentally new approach to the design and development of a high reliability point machine for the railway. Elements of the Track sub-system have been developed for this project, however they represent an incremental improvement over proven existing designs.

2 ES3200 SPECIFICATIONS

The ES3200 suite of specifications comprises 6 documents, detailed as follows:

ES3200: General Requirements for the Design and Development of Switch and Crossing Systems

ES3201: Technical Requirements for the Design and Development of Switch and Crossing systems

ES3202: Installation, Maintenance and Requirements for Switch and Crossing Systems

ES3207: Through Life Cost Model for Switch and Crossing Systems

ES3209: Configuration Management System for Switch and Crossing systems

ES3211: Special Project Documentation for the Design and Development of Switch and Crossing Systems

These specifications were developed for use as procurement specifications placing full system responsibility, including design authority, with the system supplier. They are non-prescriptive except where required for legislative or safety reasons. They were originally intended for use on "green field" sites where the designer has to consider the S&C system in its entirety. The specifications were also deemed as suitable for use on future high-speed railways with line speeds up to 255 kph (160mph).

The specification defines the key functional and performance requirements for the 5 sub-systems that make up the S&C system, namely Actuation, Locking, Detection, Track and Track Support. The specification also provides numerical targets for Reliability and Safety, stating, " the S&C system shall attain a Mean Time Between Attributable Faults (MTBAF) of at least 2 service years". One service year is defined as the Equivalent Million Gross Tonnes Per Annum (EMGTPA) of rail traffic passing over the S&C system.

3 DESIGN AND DEVELOPMENT METHODOLOGY

Industrial Actuation's approach to the development of the HPSA unit has been one of a "top-down" devolvement of requirements from a top level Functional Requirement Specification (ES3200). The classic "V" validation and verification approach was used, from which the top-level requirements were devolved into a structured design, development and test programme.

3.1 Requirements Capture

The Industrial Actuation / BBREL joint project used the ES3200 specifications as their top level Functional Requirements Specification (FRS) and Safety Requirements Specification (SRS) for the project.

To supplement this top-level specification a series of Hazard Identification (HAZID) activities were carried out as part of the safety approval process. These HAZIDs identified a number of hazards, including known areas of unreliability that lead to unsafe systems, or "men on track", associated with S&C systems. These HAZIDs proved to be a powerful tool in identifying a number of design drivers for this project. The HAZIDs, plus all other design and development work carried out, considered the full project lifecycle of the HPSA unit operating in the railway environment. This lifecycle covers the concept stage, through design, development and testing, installation and operation and finally de-commissioning.

The project also made numerous visits to S&C systems in the live railway environment and carried out a series of structured interviews with installers, maintainers and operators. This assisted with the identification of a number of key features of an S&C system, including the Point Machine, that would be both desirable for inclusion in the HPSS / HPSA design.

3.2 Devolvement of Functional Requirements into Design Requirements

One of the key definitions required in any Design Specification is that of the system and sub-system boundaries and the internal and external interfaces. This ensures that the system designer clearly understands the inputs and outputs of the system and its constituent sub-systems.

Context diagrams were produced in order to flow down the top-level requirements for the HPSA unit, to a series of sub-systems. These context diagrams, an example of which is shown in Figure 1, are a structured way in which all functionality and interfaces are captured in a simple graphical format.

Mathematical modelling of the HPSA operation was carried out as an element of the conceptual study work, to ensure that all performance parameters were identified.

At the end of this phase of the project a Functional Requirements Specification for the HPSA was produced from which a number of design concepts were generated. These concepts were evaluated by a number of techniques, notably Evaluation Matrices. One of the most important evaluation techniques was to compare every design option against a list of known problems with existing S&C systems and in particular Point Machines and Clamplocks. This ensured that any features on existing machines that were known to be unreliable or not easy to install or maintain could be designed out in the HPSA. A number of examples of this approach are detailed in Section 3.3. below.

3.3 Design for Reliability / Availability / Maintainability/Safety

The key to the reliability of any system is not only in the choice of proven discrete components, but also in the way in which these components are designed into the system, and the interaction between them. It is therefore of fundamental importance that all external and internal interfaces are defined and fully understood. This fact was to be reinforced at the commencement of the in-track trial described later in the paper.

System availability is best described as the total time in which it is fit for use, less the amount of time it is out of service due to faults or scheduled maintenance.

The overall design philosophy adopted by the project team was that the actuation / locking / detection sub-system should be virtually maintenance free. This demanded that the chosen design would require no adjustment during installation or operation, have no lubrication requirements, and have no items that require regular, or non-scheduled replacement during the system life of 25 years. A modular design approach was therefore chosen, ensuring that if a random failure occurs in service or degradation of a component is identified by Condition Monitoring, module replacement in-situ may be completed in a timely manner. System components were therefore chosen that have a proven record of high reliability in arduous operating environments.

A design philosophy is one thing, however engineering requires objective proof. Industrial Actuation's design and development programme was therefore based on their vast experience of a robust development regime involving an arduous test programme. This process allowed significant design and reliability growth to occur, without the HPSA prototype going into the railway prematurely. The history of engineering is littered with occurrences of equipment being introduced into service without sufficient development taking place, the railway being no exception.

Examples of Industrial Actuation's proven approach to designing reliability into a product may be demonstrated by a few examples used in the HPSA:

Brushless d.c. motor – use of a brushless motor means that random failures due to excessive brush wear are eliminated.

"Drive to stall", of the Switch Rails against the Stock Rails, the stall condition being detected by "motor stopped" sensors.

Switch Rail toe gap measurement is made relative to the Stock Rail thus eliminating the need for adjustment during set-up, commissioning or during operational service, as is the case with some existing point machines.

Detection of the Stock and Switch Rails is achieved using linear detectors known as Linear Variable Differential Transformers (LVDTs) that provide a linear output voltage, proportional to the relative movement between the fixed and movable rails. A common reason for "points failure" in existing Clamplock equipment is "bobbing" detection, this is due to micro-switch contact bounce caused by vibration induced by passing trains. This causes a loss of detection and thus a failure of the signalling system.

Containment of the entire drive, actuation and lock mechanism within a fully covered steel sleeper ensures that all items of equipment are protected from traffic damage such as trailing chains. This also allows the S&C system to be fully machine tampable which will permit better P-Way maintenance around the S&C system. This in turn should provide greater support for the bearers, minimise vibration and hence induced failures in the track mounted equipment, and improve ride quality through the turnout.

The steel sleeper provides good Electro-Magnetic Interference (EMI) and Electro-Magnetic Compatibility (EMC) protection as it creates a Faraday cage around the Electronic Control Unit (ECU) and the LVDTs that are contained within the HPSA.

3.4 Concept and Detail Design

Detail design using Computer Aided Design (CAD) tools was carried out, and it is not the intention of this paper to provide extensive detail of this project phase. One critical activity carried out in parallel with the detail design was an extensive stress analysis of key components, thus ensuring product integrity from a safety, functional and endurance standpoint. A number of techniques were used ranging from classical techniques through to Finite Element Analysis of key components.

A further important piece of analysis work carried out was to perform a Failure Mode, Effects and Criticality Analysis on the entire HPSA, in particular focussing on the electronics design architecture and programmable devices. These studies were useful tools in avoiding potential areas of unreliability during the design and development stages, and in predicting the Mean Time Between Failures (MTBF) for this sub-system.

3.5 Design Reviews

Throughout the design and development of the HPSA a number of structured design reviews took place. These comprised the Concept Design Review, Preliminary Design Review, Critical Design Review, and a Test Readiness Review.

The review teams comprised, as a minimum, the review chairman (usually the Project Manager), representatives of all relevant technical functions, Quality, Production, Commercial, Safety, plus Railtrack HQ's appointed technical mentor.

Each review was conducted against the requirements of a detailed checklist, and any changes required were documented in a set of comprehensive minutes. Each successive review then ensured that previous actions had been satisfactorily actioned in subsequent design and development phases.

4 QUALIFICATION TESTING

4.1 HPSA Sub-System Testing

The HPSA has completed a comprehensive set of Qualification Tests based on a Qualification Test Specification that details the tests to be carried out, the scope, duration and duty cycle(s) of that test, plus the criteria for test success.

Qualification Testing was carried out in the following key areas:

Functional Testing (Figure 2)
Endurance Testing
Axle Loading Fatigue Tests

Environmental Testing, comprising:

- Vibration Testing (Figure 3)
- Electro-Magnetic Compatibility (EMC) Testing
- Extremes of Temperature (Hot and Cold)
- Water Immersion Tests (Figure 4)
- Foreign Matter Tests (Dust and Ballast) (Figure 5)

As with any newly developed piece of equipment, fitness for purpose is an obvious but often misconceived requirement. It was an easy assumption to think that the HPSA should only move the Switch Rails from one position to another, i.e. Normal to Reverse, or vice-versa. The Qualification Test Specification was therefore compiled to ensure that considerations were made relating to a number of what-if scenarios beyond this simplistic assumption.

For example, although the Functional Requirement Specification gives an operating voltage for the Point Machine, what if this voltage is considerably more, or less than that specified? Does the system fail to operate, does it operate within allowable time limits, will the electronics withstand extremes of voltage, current etc. These considerations were applied to all tests conducted resulting in a demanding test regime. The success of any project such as this is that reliability growth is achieved during the qualification stages, such that when the product is commercially available it has been comprehensively qualified to function reliably in its operational environment.

During the test phase a number of minor test failures occurred, as soon as these happened the Company's Failure Recording and Corrective Action System (FRACAS) was used to record, analyse, and action the necessary measures to remedy the problem. This system is an extremely powerful tool for tracking problems and failures during the development phase and ensuring that corrective measures are implemented.

Testing of the HPSA was completed with the majority of the original test requirements being satisfied, whilst a number were found impossible to conduct due to the lack of railway industry data being available. Most notable was the lack of a definitive vibration specification for track mounted equipment, therefore a limited vibration test was carried out in a laboratory sufficient to provide the necessary confidence that the HPSA could undergo an in-track trial.

4.2 Off-Track Tests

The Qualification Test phase included the installation of an HPSA unit in an old turnout located off track at Yatton station, near Industrial Actuation's factory. The HPSA was powered from a Location Case powered from the local signalling system, and allowed the project to have an operational HPSA in the railway environment. Although this unit was non traffic carrying, it was put through an extended functional cycling programme that to date has completed 360,000 cycles (Switch Rail throws). This figure exceeds the number laid down in ES3200 (350,000 cycles) as being representative of 25 year service life for the S&C system.

A significant advantage of having the unit in a representative test site is that a large number of railway experts have assessed the unit in terms of functionality, installation, commissioning, maintainability, repair, and manual operation. Feedback from this peer group has either been incorporated in the trial unit at Tamworth, or has been logged and will be reviewed during the productionisation phase of the HPSA project.

Figure 6 shows the HPSA unit in the Yatton trial site.

4.3 System Integration Testing

Following completion of the HPSA sub-system qualification tests, a series of HPSS System Integration Tests were carried out to confirm full system compatibility, in terms of Form, Fit and Function. As described previously in this paper the key areas of concern relating to system reliability involved those tests of internal and external interfaces across sub-system and system boundaries.

Integration Testing was carried out at Balfour Beatty's Sandiacre site during November 1997, these tests were concluded to the satisfaction of Railtrack, their Technical Safety Assessor and Her Majesty's Railway Inspectorate.

4.4 In-Track Trial

A prototype HPSA was installed as part of a new Switch and Crossing system at Tamworth 33 points in April 1998, to commence a fully monitored in-track trial in Railtrack's West Coast Main Line (see Figures 7 & 8). This trial system is based on a conventional shallow depth (FVS24) turnout, with enhanced trackwork features and a prototype HPSA (formerly know as SAM). The prototype system is used as a trailing set of points only, this was deemed acceptable in terms of safety risk by Railtrack.

In the first 16 months of operation, over 43,000 trains passed over the system at speeds up to 170kph (105mph). The switch has operated over 18,000 times during this period without any attributable failure.

During the early stages of the trial, notification was received from the local maintenance team that the HPSA had apparently failed in service. A team was sent up to investigate and it was found that the local electricity company had severed a main power cable in the region and as a result an auxiliary supply was provided to keep the railway operational. This supply was of such poor quality that for a period of 4 hours the HPSA was being exposed to voltages and currents way in excess of those limits set in the Functional Requirements Specification, or in any Qualification Testing that had been carried out.

The FRACAS review following this failure identified that the power interfaces had not been correctly identified within the requirements specification, ES3200. Therefore, although the project had subjected the sub-system to every test required by the Qualification Test programme, it was a lack of full definition of the external interfaces that caused a failure.

Although this incident resulted in a short-term serious problem at Tamworth, it did provide useful information to allow a design enhancement prior to full-scale production of the equipment. This is a good practical example of the benefits of the in-track test site as a Reliability Growth trial for a novel system.

The production version of the equipment will be able to incorporate this and other enhancements identified through practical feedback from technical and operational issues.

5 DESCRIPTION OF THE S&C SYSTEM AT TAMWORTH

The new S&C system installed at Tamworth 33 points is based on a Railtrack standard layout, being a fully welded co-planar turnout with shallow depth switches on concrete bearers. The S&C system is designed to be "run-through" tolerant however it is not "trailable". The novel design features of the High Performance Switch Actuation (HPSA) sub-system are described below:

The HPSA unit is an in-sleeper electro-mechanical points machine providing point drive, locking and detection functions.

The HPSA is controlled by an Electronic Control Unit (ECU) which is mounted on vibration shock mounts within the sleeper. The ECU has the capability of interfacing with any type of electrical signalling system currently used within the UK railway industry and receives commands from and sends responses to the external signalling system in the same way as existing point machines. Logic devices contained within the ECU process signalling commands for a point movement and following the validation of this signal, power up and electronically commutate the motor to drive the points to their commanded position. Once the Switch Rails are locked in the correct position, as determined by a pair of linear position transducers, the ECU checks that the outputs from all position and lock sensors are acceptable when compared with factory set datum values stored in the logic devices. If these values lie within acceptable limits the ECU sends a "Good Normal" or "Good Reverse" output signal to the external signalling system, depending on the lie of the points.

Another key feature of the ECU is that it contains a register of the last 512 commanded operations of the HPSA. This register logs key performance indicators such as the time taken to complete a switch movement and the actual Switch Rail position (open and closed rails). This information may be downloaded to provide the operator and maintainer with important trend data, such that if any serious degradation of performance is noted corrective action may be implemented at the earliest opportunity and without loss of critical availability of the S&C system. This Condition Monitoring feature is considered vital for the future improvement in predictive (preventative) maintenance occurring in the operation of the railway.

The drive system utilises a dc brushless motor, which is both ac and dc immune. The main drive output is via a three-stage reduction gearbox, and rotary to linear motion is achieved via an acme-threaded leadscrew. Locking of the Switch Rails in their commanded position is via this leadscrew and an additional duplex electro-mechanical braking system mounted on the gearbox. The gearbox output actuates a central drive carriage that is connected to the Switch Rails via a pair of drive arms.

Once the HPSA commences switch operation following a command from the ECU, the motor drives the acme leadscrew that either extends or retracts depending upon the original position of the Switch Rails and the commanded operation. The Switch Rail is then driven hard up against the Stock Rail until a predetermined stall current is detected by the ECU. Once this current is detected the ECU switches off power to the motor and releases power to the dual electro-magnetic brakes which then apply a holding force to the motor output shaft, thus preventing un-commanded motion of the Switch Rails and locking them in position. The ECU receives data from both linear detectors (LVDTs) that reference the positions of the open and closed Switch Rails relative to their adjacent Stock Rails. The ECU compares the position data with that stored in the logic devices, and assuming that all values of rail position and the output from both brake detectors are within defined parameters and tolerances the ECU sends an output voltage to the external signalling system corresponding to a valid detection state. If any of the position sensors or lock sensors give an unacceptable output the ECU will give an "Out of Correspondence" signal.

7 THE WAY FORWARD

The results of the in-track trial at Tamworth, plus the results of further in-house test work carried out, coupled with end user and rail industry feedback will be used to optimise the existing prototype HPSS / HPSA design. This will result in a high performance switch actuation system that is capable of delivering fault free performance in tomorrow's railway.

REFERENCES

(1) Office of the Rail Regulator: Annual Report 1998-99

The author wishes to acknowledge the assistance provide by the following companies through their participation in this project and with the compilation of this paper:

Railtrack plc, Balfour Beatty Rail Limited, WS Atkins Rail Limited, Halcrow Transmark

FIGURES

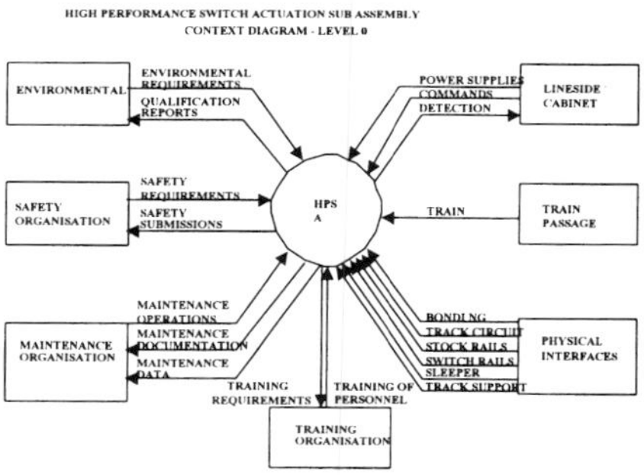

Figure 1: Typical Context Diagram

Figure 2: Endurance Test Rig

Figure 3: Vibration Test Rig

Figure 4: Water Submersion Tests

Figure 5: Yatton Trial Site

Figure 6: Tamworth Trial Site

C570/013/99

Stainless steel – an innovative answer

J M HAYNES
Civil Engineered Products Limited, Rotherham, UK

Key words : Stainless Steel, Reinforcing Bar, Tie Bar, Corrosion, Civil Engineering, Structures, Repair, Maintenance

SYNOPSIS

Stainless steel is a relatively new material. The use of stainless steel in bar form, both as a reinforcing bar and tie bar, is in its infancy. The virtues of stainless for these applications are abundant. With regard to fault free structures, the technical advantages of stainless steel are proven and infinite life is predictable. In contrast the corrosion of conventional carbon steel reinforcing bar is causing serious problems in existing RC structures.

The current restraint to greater usage of stainless steel is the perceived high cost. A new technique of providing reinforcing ribs and threads on stainless steel round bar is described. The end result is a highly versatile, lower cost stainless steel product. This is confirmed by weight savings of up to 40%, compared with conventionally machined stainless steel bar products

1. INTRODUCTION

In engineering terms stainless steel is a relatively new product. Early experiments with high chrome bearing irons did not lead to commercial applications. Harry Brearley discovered stainless steel in 1910 and the first arc furnace cast was made in Sheffield in 1913. Harry Brearley was subsequently recognised as the inventor of stainless steel and he was the first to recognise the commercial implications of such a material. The combination of Sheffield's steelmaking, and its cutlery industry led to the first utilisation of stainless steel. This usage spread into the American cutlery industry and led to refinements of the steel until by the 1920's a wider range of new grades had been developed, although they were expensive. Since then the development of stainless steel has been biased towards the applications and manufacturing aspects rather than the basic chemical composition. The prime ingredients for corrosion resistance are chromium and nickel.

The use of stainless steel is now common in many branches of engineering and across a large number of products. Even mundane items such as motor vehicle exhaust systems are now heavily reliant on stainless steel for their longevity. Domestic examples abound, washing machines, dish washers, kitchen utensils, working surfaces and flue liners give some indication of the development of stainless steel applications. Industrial usage is manifold in the process industries, pharmaceutical and power generation plant, marine and medical equipment. With regard to buildings and structures the application of stainless steel has been primarily for architectural requirements.

2. OBSERVATIONS

All the above applications have a predominant bias towards using stainless steel in the flat product form of strip, sheet or plate.

One of the exceptions to the above observation was that in the mid 1960's two companies, United Steel and George Clark, started a joint product development programme to produce a stainless steel high strength ribbed bar which was based on the then recently developed ribbed high yield carbon steel reinforcing bars. The stainless version was used in small quantities and for special circumstances. At that time, the usage of ribbed carbon steel reinforcing bar was going through a tremendous growth and it is not unreasonable to say was incorporated in all major civil engineering constructions. This newly found method of construction became very fashionable to the detriment of conventional structural steel applications.

Ironically, one of the last bastions of conventional steel structures were the railways and today when driving on a motorway wherever you see a steel girder type bridge it will be a railway bridge. This reflected the reluctance at that time to use reinforced concrete because it was unproven. It would be interesting to know whether the insistence on remaining with steel rather than reinforced concrete was a premonition of the forthcoming problems or simply an example of railway engineering conservatism. Whatever the cause the resistance has probably saved thousands if not millions of pounds of remedial costs. By contrast, the highway authorities on the same motorways are now facing ever increasing demands for funding remedial work caused by corroding reinforcing bar. The USA Highway Authorities are experiencing serious problems of corrosion in bridges with a current estimate of 600 000 road bridges in distress and a repair bill estimated in 1998 at $200 billion. It is sad to say that not one of those bridges achieved anything like the design life envisaged and the repair cost is now said to be four times the original cost of construction (Ref. 1). Fig.1 shows a typical corrosion induced hole in a USA bridge over rail track and Fig.2 shows the corrosion damage to carbon steel reinforced pre-cast units in a multi-storey car park.

Fig.1
Bridge Deck Failure due to Reinforcing Bar Corrosion

Fig.2
Car Park Walls Damaged by Reinforcing Bar Corrosion

There is no doubt in the mind of most engineers associated with conventionally reinforced concrete structures that the corrosion problem means they are dealing with a "ticking bomb". Their designs have, or will, fall well short of being fault free structures.

In contrast there is sufficient evidence to confirm the longevity of stainless steel bar in civil/structural applications. A recent publication issued by the US Department of Transportation - Federal Highway Administration (Ref 2) reports observations on US bridge decks which utilised stainless steel reinforcement. Of the two cases involved, both show no significant corrosion and the stainless steel rebar appeared to be largely unaffected by placement in concrete in applications with up to 16 years service. These results were obtained by taking core samples from the bridge decks. Those results endorse the prediction that infinite life is available from structures incorporating stainless steel bars.

The above observations have two important messages with regard to achieving a fault free infrastructure. Firstly, the use of conventional carbon steel reinforcement is fraught with risk. There is now clear evidence that where the rebar becomes exposed to carbonation from the atmosphere and, even worse, chlorides, the products of corrosion on the bar are sufficient to burst and dislodge the adjacent concrete substrate. This risk of attack, often assisted by inadequate cover, shrinkage cracks, settlement cracks, mechanical damage and porosity of the concrete makes conventional reinforcing bar a poor choice for long life and fault free structures. Secondly, the case for stainless steel reinforcing bar is proven and it can be used with confidence in the most aggressive conditions to achieve long design lives to the point of forecasting infinite life in terms of constructional products. These simple observations raise the question, "Why isn't stainless steel used more extensively ?".

3. CONSTRUCTIONAL STAINLESS STEEL

Within the grades of stainless steel referred to in previous paragraphs two compositions, both of the Austenitic type, stand out as being the most commonly used for civil/structural engineering applications. These are Grade 304 (18-8) and for higher resistance in extremely corrosive situations Grade 316 (17-11-2). Both these steels are manufactured on a wide scale and can have mechanical properties equal to or greater than conventional high yield reinforcing bar. They both offer an opportunity to use them as direct replacement for designs based on conventional carbon steel reinforcing bar. The concrete Society have recently issued a Technical Report (Ref 3) which contains a comprehensive résumé of stainless steel reinforcement

Of the other stainless steels there are occasions when very high strength is required and this is catered for by using precipitation hardened steels (17-4) or (15-5). However, for the purpose of this Paper there is no further reference to using higher tensile strength grades..

So far, consideration has been given to recognising faults in relatively new structures and a technical solution based on using stainless steel. Railway engineers face additional problems because many existing structures are based on masonry operating in a totally compressive mode i.e. pre-reinforced concrete structures. Ironically, most of them have survived significantly better than would now be predicted for their more modern counterparts.

However, they have exceeded their original design life and been subjected to higher loadings than was originally intended. A common cause of distress in these structures is their inadequacies in dealing with tensile forces and changes in geometry due to settlement. A common example of this problem is the outward movement of spandrel walls on bridges and viaducts.

The traditional answer to this problem is the use of post installed tensile reinforcement often in the form of tie bars and anchors, combined with pattress plates. In early times the engineer had limited choice of tendon materials and there was extensive use of plain carbon steels which were often applied externally to give the necessary tensile restraining forces.

The utilisation of stainless steel, has grown significantly in recent years and is now a well established method of applying a post reinforcement system. Just as stainless steel offers a fault free structure in the new build situation it also offers a fault free structure where repair and refurbishment is being considered. Moreover, the utilisation of stainless steels in refurbishment work of this type usually represents a very small cost relative to the structure as a whole. So much so that it begs the question, "Why do engineers continue to use carbon steels for these applications ?".

To summarise, problems with conventional reinforcement have been established, stainless steel is a proven solution, repair and refurbishment is similarly enhanced by using stainless steel and yet its application remains limited.

There are three perceived restrictions to greater usage of stainless steel. Firstly cost, secondly lack of product knowledge and design confidence and thirdly availability.

4. DESIGN EXAMPLES

Substitution of stainless steel for carbon steel reinforcement is a simple operation and there are no constructional applications where this cannot be undertaken. Both stainless steel and carbon steel reinforcement are made to have the same mechanical properties and although they have separate British Standards, (Ref 4 & 5), those standards could be easily combined to make the only difference one of material composition. Fig.3 shows an example where direct substitution was used in a bridge deck and this was one of the bridges referred to in Ref. 2 which reported the excellent long term performance.

Fig. 3 Substitution for Carbon Steel
Placing Stainless Steel Reinforcing Bar in a USA Bridge Deck

Wherever long life is required stainless steel should be first choice, e.g. structures where usage will continue for the foreseeable future such as. historical buildings and strategic structures. Fig. 4 shows Ely Cathedral which has repair ring beams in the towers which utilises stainless steel reinforcing bar. Although not a practical consideration it is fact that just as the stone from Ely Cathedral would be reusable in say 1000 years' time, so would the stainless steel reinforcing bar.

Fig. 4 Infinite Life Requirement
Remedial Strengthening with Stainless Steel Reinforcing Bar for Ely Cathedral

Wherever the environment is aggressive, e.g. exposure to road salts, marine locations, areas difficult to protect against mechanical damage, in structures with voids and unknown infill, in rock and structures with fissures and cracks that permit moisture ingress, then stainless steels should be first choice. Fig 5 shows Scarborough Spa where an existing marine structure was extensively rebuilt using stainless steel both in the precast and the in situ concrete. In this case the selection of materials was determined by the anticipated degree of exposure to sea water chlorides. The precast units directly exposed to sea spray used 316 Grade steel, other units used 304 Grade steel and the in situ ground beams used a combination of 316, 304 and eventually carbon steel as they progressively moved in land and into dry conditions. Fig 6 shows the pre-cast units during manufacture.

Fig. 5 Marine Exposure
Extensive Reconstruction Using Stainless Steel Reinforcement at Scarborough Spa.

Fig. 6 Precast Units
Placing Stainless Steel Reinforcing Bar in Moulds for Precast Units.

Wherever the cost of repair would be high in terms of subsequent access, e.g. in terms of say scaffolding, disruption to traffic, submerged in water or in tunnels then stainless steel should be first choice. Fig. 7 shows the Thames Wall at Wapping, which is precast brick faced concrete panels incorporating stainless steel reinforcement. The panels were designed to look like brick to match existing surrounding buildings.

Fig. 7 Maintenance Free Infrastructure
Precast Brick Faced Panels with Stainless steel Reinforcement and Fixing System at Thames Wall, Wapping.

The technical case for stainless steel in civil/structural engineering is so strong that if its cost were similar to carbon steel the usage of that latter product would be restricted to very short term applications. However, the reality is a significant increase in the cost of stainless steel, between 5 to 8 times that of carbon steel. This difference is partly explained by the higher value of metals involved in manufacturing stainless steel.

When life cycle costing is adopted there is a point where stainless steel reinforcing bar becomes cost effective. Obviously, the closer stainless steel cost is to mild steel cost the shorter is the design life required to validate stainless steel as first choice. When this is combined with a selective use of stainless in high corrosion risk areas and certainly where design lives in excess of 80 years are involved, on the evidence to date, the case for stainless is already proven. A recent example of this is given in a case study (Ref 6) concerned with a bridge over the Rhine at Schaffhaussen in Switzerland. Even with stainless steel costing 10 times more than carbon steel reinforcement, the total initial cost increase was only 0.5% resulting in a life cycle cost saving of 13%, combined with a bonus of minimum maintenance and traffic disruption.

There are two other factors which require addressing by both the manufacturer and customer. If a stainless steel reinforcing bar manufacturer adopted a similar manufacturing mentality to a carbon steel reinforcing bar manufacturer there would be a significant price reduction to, say, something less than £1000 per ton for a 304 Austenitic steel reinforcing bar. The customer also compounds the current high price mentality by accepting the manufacturers prices and becoming discouraged to the point of being restrictive in applying stainless steel reinforcing bar. With time these compounding negative factors will diminish and one day an enlightened manufacturer will grasp the opportunity to satisfy an obvious market demand and/or an engineer customer will create sufficient demand that it deserves the right price.

5. AN INNOVATIVE ANSWER

Civil Engineered Products Limited is a company specialising in bar products, both stainless and carbon steel, utilised in civil and structural applications. The previous text reflects the company's own experience of a market demand not being met by the supplier. This shortfall was the driving force towards finding alternative ways of producing stainless steel bar products which overcame some of the perceived difficulties.

A recent innovation, Rotherbar, offers designers bar based stainless steel products at significantly lower costs, combined with the ready ability to substitute stainless steel for carbon steel designs. This product was a finalist in the Construction News Awards for R & D (Materials) 1999 and was designated a Millennium Product by the Design Council in April 1999 (Ref 7)

The patented technique is to produce ribs and threads on plain bar by precision weld deposit resulting in savings of up to 40%. Fig. 8 illustrates ribs which have been weld deposited on to a bar. The technique has been referred to as positive machining because it puts metal on rather than machining it off.

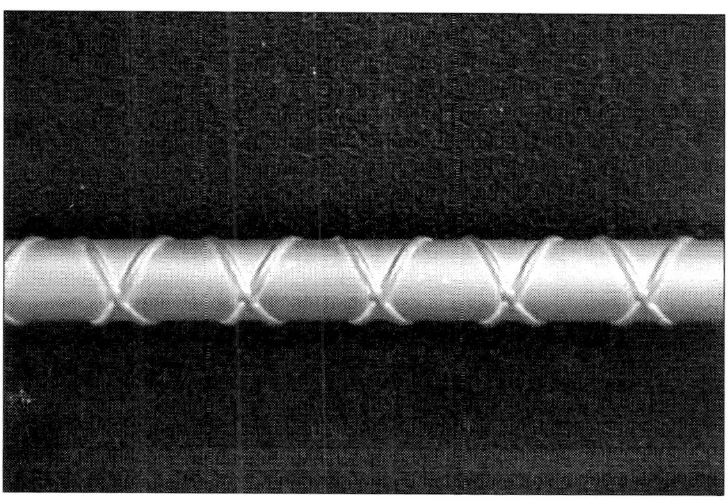

Fig. 8 Rotherbar
Reinforcing Bar with Weld Deposited Ribs.

Fig. 9 illustrates a fixing bolt made from conventional stainless steel rebar with a machine threaded end, compared with a Rotherbar threaded end with appropriate reinforcing ribs at the other end. Both bars have identical strength which is determined by the thread root diameter. The saving in material is clearly illustrated.

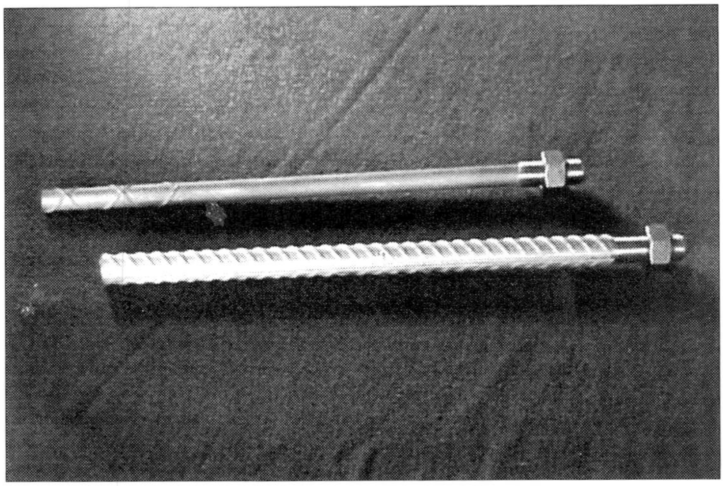

**Fig. 9 Stainless Steel Fixing/Anchor Bolt
Rotherbar v Conventional Rebar
Showing the Weight Saving by Utilising Weld Deposited Ribs and Threads.**

As well as producing tie bars and reinforcing bars in the conventional sense the process allows almost infinite variability in terms of combining threads, plain bar and ribs all on to one product and in whatever order they are required.

The above process and the resultant products have been developed and tested in collaboration with Sheffield Hallam University. World patents are in process and the manufacturing system allows a highly portable manufacturing technique that could be "on site" if demand was of sufficient magnitude.

6. CONCLUSIONS

The technical case for stainless steel in new construction is so strong that when the price becomes closer to carbon steel its usage will become common place.

If life cycle costing is used at the planning stage the case for stainless is good even with the current relatively high initial cost. Moreover, if the design is refined and utilises stainless steel in the areas prone to corrosion, combined with lower corrosion resistant reinforcing bar in safer areas, the economical case for using stainless steel is readily proven.

In structural repair, where the quantity and cost of repair materials is usually modest compared with the original structure, the case for stainless steel is so strong it raises questions as to why engineers continue using short life materials.

The perceived high cost of stainless steel is the main barrier to greater utilisation. New innovative products are reducing those costs, as will greater usage of conventionally produced products.

There is a marketing/pricing aspect to stainless steel reinforcing bar that results in higher than need be prices. A closer relationship and change in attitude by both manufacturers and users could bring about significant economies.

With regard to the need for "fault free infrastructure" stainless steel should be first choice.

REFERENCES

1. Nickel, September 1998 Vol 14, Number 1, pp 12-15

2. US Department of Transportation - Federal Highway Administration. Publication **No** FHWA-RD-96-085, October 96. The Corrosion Performance of Inorganic - Ceramic, and Metallic Clad Reinforcing Bats and Solid Metallic Reinforcing Bars in Accelerated screening tests.

3. Concrete Society Technical Report No 51, 1998. "Guidance on the Use of Stainless Steel Reinforcement". ISBN 0946691 89 X

4. BS 6744, Austenitic Stainless steel bars for the reinforcement of concrete.

5. BS 4449, carbon steel bars for the reinforcement of concrete.

6. EURO INOX Case study Life Cycle Costing and Stainless steels - River Crossing Highway Bridge, July 1997

7. Design Council. Millennium Products , 34 Bow Street, London WC2E 7DL

C570/025/99

Paying the price – is performance contracting the answer?

G R H HAMPSHIRE
Booz • Allen & Hamilton, London, UK

ABSTRACT

There are many initiatives to improving the performance of railway infrastructure by reducing the number of faults. Recent restructuring of the railway industry in some countries has facilitated innovative performance contracts between infrastructure owners and their contractors. This paper considers how performance within a railway context can be defined and how different system characteristics can be considered. The use of performance contracts for the construction, upgrading, renewal and maintenance of rail fixed infrastructure around the world is reviewed. Observations are made about the success of the use of performance contracts in improving the service delivery of the railway system. The initial indications are that some improvements have resulted in the short term.

1. INTRODUCTION

The railway industry has undergone fundamental changes over the last few years, driven by the need for improved levels of service, increased performance and greater reliability. In this context, the restructuring of many railway organisations has resulted in the separation of infrastructure provision and train operations, and the rate of change seems set to increase. This vertical disintegration allows greater visibility on the contribution of each component of the railway system to the delivery of the service to the customer. In particular, it has highlighted the contribution of the fixed infrastructure to the delivery of a safe, reliable and convenient service.

Infrastructure is one part of the railway system which, for the purposes of this paper, can be represented by five activities (Figure 1). Infrastructure is built, maintained and renewed, either by direct labour or by contractors. Infrastructure is then made available to a network control organisation which provides train paths to train operators. In turn, these train operators provide services to users (both passenger and freight).

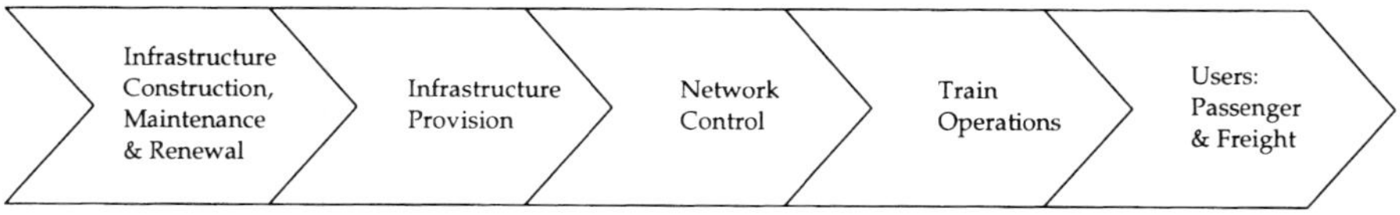

Figure 1: The Railway System

The outsourcing to contractors of activities once undertaken in-house is big business in many industries and is increasing. This is true of the rail infrastructure industry (Table 1) where the current trend is to increase the use of contractors, rather than utilise direct labour. However, there is still a wide variety of approaches.

Table 1: Contracting Activity by Country

	Track				Signals
Country	Renewals	Periodic	Routine	Inspection	
Finland	●	●	●	●	●
UK	●	●	●	●	●
France	●	●	●	○	●
Germany	●	●	●	○	●
Netherlands	●	●	●	○	●
Spain	●	●	○	○	●
Sweden	●	●	○	○	●
USA	●	●	○	○	●
Austria	●	●	○	○	○
Australia	◔	◔	◔	◔	◔
Denmark	●	○	○	○	○
Portugal	●	○	○	○	○
Norway	○	○	○	○	○
Switzerland	○	○	○	○	○

↑ Increased outsourcing

● *By contract* ◔ *Partially by contract* ○ *In-house*

The increased visibility of the infrastructure contribution has encouraged the use of innovative contracting arrangements. One such arrangement is the use of performance contracts where a portion of the contractor's renumeration is linked to its performance against certain targets.

Has this trend helped to improve the management of railway infrastructure and improve service delivery by reduced impact of failures or simply introduced new problems? Is the concentration on fault reduction to improve service delivery in the best long term interests of the industry? This paper considers what is meant by performance and how it can be defined for infrastructure, how this is translated into contracts for infrastructure services and the experience with performance contracting to date with particular respect to overall system performance.

2. DEFINING PERFORMANCE

The crucial element of any performance contract is the performance specification. At the highest level, this requires a definition of the business purpose, or for the railway, the system purpose. However, many businesses, let alone railways, are unable to set out their business purpose. In a recent Fortune 500 survey in USA, over 75% of respondents did not have a clearly defined set of long term goals! The rail industry should take heart that the journey to the well defined business objective has actually started, with the development of measures such as the "delay minute".

The use of performance contracts can assist both infrastructure providers and contractors. The infrastructure provider has to define the key measures of performance which require consideration of both the needs of the infrastructure and of the end users. When incorporated into a performance specification, the contractor is then by definition forced to focus on the same issues. The end result should be a shared focus on the ultimate end user by both infrastructure provider and contractor.

Some work has been done to date on what constitutes performance in a railway context such as the delay minute but more work requires to be done. In the eyes of the users, railway performance can be considered at two levels (Table 2):

Table 2: Railway Performance – the Customer's View:

Criteria	Definition
Meeting initial expectations	Delivery of the promised level of service i.e. reliability and punctuality
Enhancing expectations	Either improved flexibility (more station stops) or reduced journey time

However, such performance measures may be of secondary issue to shareholders in the privatised railways where financial indicators such as the share price may be of more interest.

3. INFRASTRUCTURE PERFORMANCE ISSUES

Utilising performance specifications to define the system and services to be delivered under the asset related contracts introduces several challenges:

- Inter-relationship of performance requirements
- Measurement of Performance
- Interface control
- Access to information
- Asset condition

Inter-relationship of performance requirements

Many performance measures are inter-related. For example:

- availability can defined as the relationship between frequency of failure and the time to repair the failure (corrective maintenance)
- maintainability can be defined as the relationship between maintenance frequency and the time to perform maintenance
- reliability can be defined as the frequency and/or impact of failure

It is essential within output specifications to ensure that the overall performance goals of the system are acknowledged when targets set for inter-related performance measures are combined.

Measurement of Performance

All requirements of the output specification must be measurable and/or testable. Thus, performance measures, and the tools to be used to confirm the achievement of performance targets, will need to be clearly defined for each phase of the contract. In practice, the performance indicators are unlikely to be perfect and it requires the co-operation of both parties to improve the indicators by focusing on the those that help in achieving pre-defined goals. For instance, Railtrack in the U.K. have committed to certain levels of track quality utilising the output of track recording vehicles as a proxy for such quality.

Interface control

In order for contractors to take full responsibility for the delivery of performance, they must be able to control a majority of the parameters which influence the performance measure. Thus, the boundaries of control and assumptions which should be made regarding interfaces must be clearly defined.

Access to information

The accessibility of appropriate data is a major consideration in defining the performance measures. It will be important that the performance data required is easily accessible, objective, reliable and inexpensive to collate. It is important that asset information is available to both parties, regardless of who actually collects it.

Asset condition

Assets can be grouped into three categories for the purpose of determining their condition in normal operating mode:

- assets whose condition directly effects network performance, e.g. signalling
- assets whose condition has little or no effect on performance, e.g. earthworks
- assets whose condition is partially reflected in network performance, e.g. track

To be an effective performance measure, 'asset condition' must be clearly defined. Factors such as fitness for purpose, projected asset life and performance, refurbishment and renewal criteria and safety integrity are all possible definitions. A key issue is how to specify long term condition of the infrastructure whilst considering short term impacts on train performance.

4. WHY CONSIDER PERFORMANCE CONTRACTS?

To date, the prescriptive form of contract has been the most common. The standard forms of contract such as I.C.E. (Institution of Civil Engineers) and F.I.D.I.C. (Federation Internationale des Ingenieurs-Conseils) require the specification by the employing organisation of what it wants delivered and it is the employer that takes the risk that the delivered product will achieve the required levels of performance. Such contracts are clear commissions with a clear start and finish and a tangible end product. The contracting process is well understood and there is much experience of the process with a concentration on price and contractual performance. However, it relies on the employer to complete the feedback loop by communicating the in-service performance of the new asset to the product specifiers. It is doubtful if such feedback was encouraged given the split between "New Works" and "Maintenance" departments of the traditional railway organisation.

In contrast, performance based contracting is characterised by:

- a specification of outcomes leaving the contractor free to choose the most appropriate method
- performance related payments
- shared risks with transfer to the contractor of technical, programme and financial risk
- management throughout the lifecycle
- better utilisation of contractor knowledge
- increased transparency

The key difference between the two types of contract is that of risk transfer. In the traditional contract, the employer retains the risk of performance by defining inputs but has greater control over long life issues. Under a performance contract, the contractor is responsible for all outputs but the employer has little control over inputs but relies on a more mature relationship between the two parties.

5. PERFORMANCE CONTRACTS IN PRACTICE: THE GLOBAL PERSPECTIVE

This section considers the extent to which performance contracts have been used for new construction, renewals and maintenance and the results that have been achieved.

New construction

The use of performance contracts for the construction of infrastructure is only just commencing, aided by the advent of private financing of infrastructure. Three examples of its use are

- Channel Tunnel Rail Link
- HSL Zuid, Holland
- Hudson Bergen LRT, USA

The Channel Tunnel Rail Link project has brought much best practice together for the procurement process by the use of design and build contracts and various other incentives. The project was originally conceived as a vertically integrated railway with performance of all assets within the rail system linked to the income in terms of receipts to London and Continental Railways. Now the train operators, namely Eurostar, will pay a track access charge to the new owner, Railtrack and Railtrack will bear the risk of performance of the asset. The opportunity to incorporate maintenance with new build was not taken.

In the Netherlands, the innovative use of private capital to finance the building of the infrastructure systems element of the new high speed line between Amsterdam and Belgium (Figure 2) has enabled the use of performance contracts to be utilised for the first time for new infrastructure. The structure of the performance scheme, developed initially by Booz·Allen & Hamilton (BAH), is still under discussion but will include elements of:

- impacts of infrastructure on train delays and cancellations
- the attainment of specified asset condition measures
- the ability of the system to deliver the required system capacity

Figure 2: HSL South Route Map

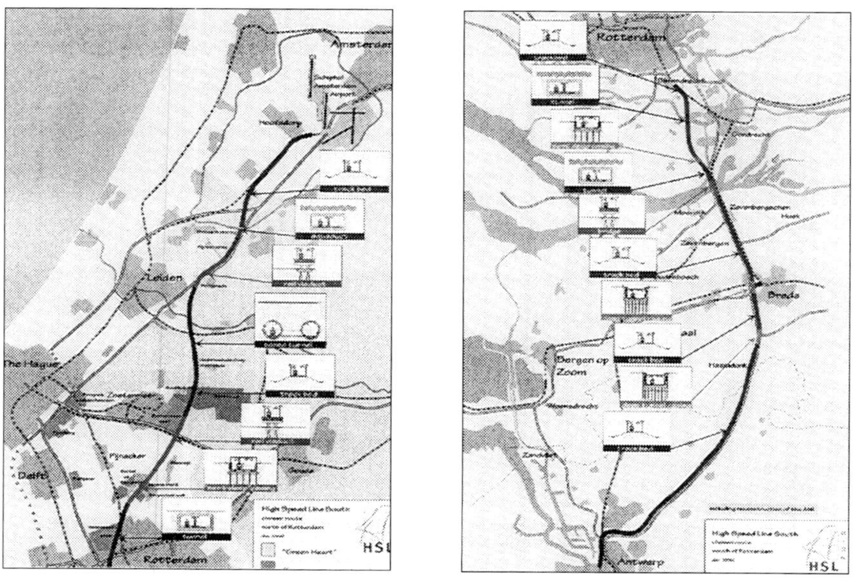

This process is currently in the procurement phase but much attention will be focused on its ability to deliver value for money and it presents a challenge to the private sector to deliver infrastructure performance at the right price. The Netherlands Government believe that such a challenge is attainable over the 30 year concession period.

In the USA, a DBOM (design, build, operate and maintain) contract is being used to construct the Hudson Bergen Light Rail Transit System. Initially this is a 10 mile, two-branch, 16 station light rail line along the west side of the Hudson River from Bayonne and Jersey City to

Hoboken. An existing rail line will provide connections to New York City, on the opposite side of the Hudson River. Later extensions to the north and south will expand the system to 21 miles and 32 stations. The contract includes 15 years operation and maintenance. Initial indications are that the construction period has been reduced compared with conventional prescriptive contracts and that whole life costs over the operating period will be lower.

Infrastructure Renewal

An example of the use of performance contracts in the renewal phase is the Docklands Light Railway (DLR) in London. In the early 1990s the performance of the DLR was giving much cause for concern with system availability as low as 81%. A number of discrete prescriptive contracts for vehicle enhancements and infrastructure upgrades were let but it was clear that these contracts on their own would not deliver the required results. BAH, in a joint venture, took over these contracts under a Prime Contractor organisation with payment based on the delivery of a specified level of performance, rather than specified products. This required the demonstration over an extended period of time of pre-set levels of availability. The consortium was able to demonstrate an improvement of availability of over 16%, delivering 97% system service availability. Despite its success, this performance method of contracting has not been utilised on such a scale elsewhere in UK railways.

There is a notable absence of the use of performance contracts in the field of track renewal, both in UK and elsewhere. This activity is a significant proportion of the infrastructure spend of most railways yet procurement of this service is undertaken on almost entirely prescriptive means. The continuation of such a division between installation and maintenance, has prevented a thorough understanding of the whole life cost issues associated with track that are currently much in vogue.

Maintenance

Performance contracting has had most impact in the area of maintenance, but this is the area where cost information is least understood. The amount actually spent on infrastructure maintenance is well known, but not the amount that should be spent. There is a complex interaction of maintenance cost drivers such as traffic type, speed, axle loads, inspection frequency, condition, renewal policy, etc. The current debate on the fixed and variable component of track access charges in the UK illustrates the nature of the problem.

Railtrack is the most aggressive user of contractors of all the railway infrastructure owners and operators. All its maintenance work is undertaken by contract. Currently this work is being re-tendered against a revised form of contract but the original maintenance contract, the RT1A contract has had interesting results. These contracts were due to last for between 5 and 7 years and had some performance components. At the outset, it was envisaged that there would be three elements to the performance scheme:

- train delay
- track quality
- safety

In the event, it proved difficult to define a safety performance scheme, based on how the indicators of such a scheme would be detailed. The train delay scheme has generally been viewed as a success, whilst the track quality scheme has led to a situation where only in the year 2001 will track quality return to pre-privatisation levels. Furthermore, there is no measure of longer term issues such as infrastructure condition or asset life. Whilst this example sheds some light on performance contracts, it must be remembered that the original RT1A contracts had several purposes, one of which was to facilitate the sale of the UK rail industry. The amount of performance based payment was restricted to 5% of the annual payments which might represent the contractor's profit, but under-performance would not result in financial ruin. The new Railtrack maintenance contracts will include a new performance regime with benchmarks which will rise each year. Contractors will therefore have to achieve year-on-year performance improvement in order to maintain the present level of income from the performance regimes.

In Australia, the Rail Access Corporation (RAC) of New South Wales (NSW) have embarked on an ambitious outsourcing of infrastructure maintenance. Using an open book alliance method of contracting, the contractor is fully re-imbursed his costs. However, the financial performance of the contract in terms of profit is related to weighted performance criteria. These performance criteria are both process and output related. The process related criteria which account for some 75% of the performance fee are:

- safety
- progress against agreed contract targets
- costs against target

However, it is clear that such measures can equally be used for prescriptive contracts. The output related criteria are reliability (basically possession over-runs) and availability. The performance measures are targets which are agreed on an annual basis such that there is a steady improvement of performance over time. The initial results are encouraging: savings in maintenance costs have enabled access charges to operators to be reduced by 20% and infrastructure incident rates have fallen (Table 3), achieved against a background of increased traffic.

	1992-5	**1996-9**
Civil engineering	7.8	4.5
Track	5.2	3.6
Misalignments	1.2	1.2
Broken rail	2.1	.6
Signals	1.3	.9

Table 3: RAC, NSW: Average infrastructure incidents per month

In Europe, NS Railinfrabeheer (RIB), the Dutch rail infrastructure company is pursuing a transition approach to contracting rather the big bang approach adopted by Railtrack and RAC. It was decided in 1994 to manage infrastructure maintenance and repair on a contracted out basis using performance contracts. During the period to 1998, input activity based contracts were negotiated with existing contractors so that both parties could gather the information necessary for the successful introduction of full performance based contracting. The process of letting these contracts in full competition has recently commenced and should result in 37 output specified contracts being let by 2002. These contracts are envisaged to last

5 years, with annual variations on the basis of the previous year's performance. However, the first contracts will only "go live" in 2001.

6. OBSERVATIONS

Whilst the use of performance based contracting is at an early stage and has concentrated on maintenance activities, it is possible to undertake some analysis of experience to date.

Type of work

The relative importance of type of work can be considered against their relevant contribution of each to life cycle costs. Generalisations are difficult but the following is suggested:

Construction:	70%
Maintenance:	15%
Renewal:	15%
Total life cycle costs:	100%

Performance contracts are becoming common for maintenance but this activity is a relatively low contribution to life cycle costs. However, despite the success of the DLR Prime Contract, there appears to be a reluctance to use similar performance contracts for infrastructure renewal. Similarly, the introduction of a performance based contract for the new build type of contract is only being considered in The Netherlands.

Contract length

Most maintenance contracts are placed for less than seven years. The industry average is about five years. This is approximately half the expected life of the short-lived signalling systems and more like 10% of some of the longer live assets such as track and electrification. In contrast, DBOM contracts for new construction are longer and in the case of private finance, can last for over 25 years.

Annual review of targets

Both RAC and RIB have annual reviews of performance to take account the experience of the previous year. Railtrack have incorporated such a feature in their latest contracts but had no such review with the RT1A contract.

Costs

The indications of the process in Australia and UK are that maintenance costs can be reduced by the use of performance contracts – RIB in The Netherlands expect similar savings. This is also true for new construction were significant savings in overall project costs appear likely.

Long term trends in infrastructure condition

Considering the discussion above about the need for infrastructure condition information, it is surprising that none of the maintenance contracts considered above have any mention of long term trends in condition. DBOM contracts generally include a definition of asset condition at the end of the contract but not during its lifetime.

Train performance

In the UK, the number of faults recorded on a network-wide basis has not varied significantly since the introduction of performance contracts but there has been a reduction in delay minutes of some 10%. This suggests that the performance contract has improved reaction to the effects of failures rather than decreasing the number of failures.

7. CONCLUSIONS

The restructuring of the rail industry has facilitated greater transparency of the contribution of the infrastructure to the performance of train service delivery. The use of performance contracts for rail infrastructure is increasing but is still at an early stage. Whilst there is limited experience of actual performance compared to that expected, there are some indications of likely outcomes.

There is a tendency to concentrate the use of performance contracts on maintenance activities. This could send the wrong message of requiring short term gains at the expense of long term performance. Initial indications in the UK and Australia are that the adverse impact of infrastructure failures on train performance has been reduced. In the UK, this improvement has been by improved response times whilst the number of incidents has been reduced in Australia.

The experience of one railway renewal using performance based contracting has been favourable and its use should be extended. However, there is little use of performance contracting for infrastructure construction which has greatest contribution to life cycle costs.

The initial indications are that performance contracts do bring benefits in terms of costs and infrastructure performance. It is yet to be seen whether this method of contracting will deliver the level of performance improvements required by all users in the long term. The use of full life contracts such as that in The Netherlands on the HSL Project and the Prime Contract role may point to the way forward.

Reliability Above and Below

C570/008/99

High-reliability electrification equipment

A BRADWELL
Consultant, Derby, UK

SYNOPSIS

The paper describes the failure mechanisms of railway electrification infrastructure equipment, based on the author's 30 years experience of investigating faults on the UK railway 25kV overhead lines and 750V 3rd rail networks. It describes monitoring techniques which can be used to check the condition of electrical equipment on site, both manually or automatically, to improve reliability and safety. It concludes by detailing areas where reliability can be improved by new design and better monitoring.

1 INTRODUCTION

Electrical power supply is an essential part of a modern railway. The overhead line / 3rd rail infrastructures and the electrical power supplies form a centralised system, without which no trains can run. This imposes a reliability burden on the electrification equipment which is far higher than that of individually powered trains, such as diesels, where the failure of a single unit in service does not disrupt the entire service. Indeed the reliability needs to be higher than that of track and signalling equipment, where alternative track or routes are available should one track be damaged.

The author has been involved for many years in assessing the reliability of electrification equipment on the old British Rail (BR) infrastructure in the UK, now owned by Railtrack. The failure statistics are based on experience, linked to data collected by the old BR HQ and Regional Electrification Engineers, but only broad figures are given (Figure 1).

2 HOW IS THIS HIGH RELIABILITY OBTAINED?

Experience in the Electrical Distribution Industry, particularly the Electricity Boards (1) shows that electrical high voltage components provide long lives with minimum maintenance. Transformers, switchgear, cables, overhead line (OHL) components all had lives in excess of 40 years, with most of the system having been installed in the 1950-60's. The same is true of railway electrification equipment, with the bulk of the old BR Southern Region (now Railtrack South Zone) 3rd rail 33kV system and the West Coast Main Line (WCML), Mk1 Scotland and Anglia

25kV OHL systems having been installed in the 1950-60's. Provided the electrical rating of the equipment has not needed to be increased due to increased service performance, the electrical equipment will perform its duty for over 40 yr. The high reliability stems from the lack of moving parts in electrical equipment, an understanding of the failure mechanisms and a conservative approach to design and testing; high voltage 25kV equipment is designed to withstand peak voltages of 200-250kV and is 1 minute tested to 95kV, giving safety factors of 4-10.

The overhead wires and 3rd rail parts of the infrastructure suffer much higher unreliability because of the large expanse of equipment exposed to the environment and to vandalism and because of the moving pantographs and shoegear.

3 WHAT HAPPENS WHEN ELECTRIFICATION EQUIPMENT FAILS?

During this 40 year life there will be, and have been, failures of equipment. When high voltage equipment (usually the insulation) fails, the fault current path of 10,000 - 100,000 Amps is so energetic that steel tanks, porcelain insulators and cable sheaths cannot contain the pressure and an explosion occurs. If the fault is in air across an insulator on the OHL or 3rd rail, a power arc flashover occurs, but the equipment is rarely damaged and the power can be re-instated quickly. Such flashovers, due to birds and debris, make up over 50% of all fault trippings. The power supply control system (SCADA) is designed to control the fault by switching off the power within 1 second, and limiting this loss of power to the vicinity of the fault. The system also allows the fault to be isolated, by remote switching, to enable power to be re-instated by another route, whilst the faulty equipment is repaired. This applies to the trackside power supplies, but if the 3rd rail or overhead line is damaged, no trains can pass until the fault is repaired. The OHL is easily damaged by a broken pantograph and difficult and slow to re-instate. The pantograph itself may be damaged by blown debris or vandalism debris or by failure of a number of OHL components or insulators, which can hang down. The broken pantograph will then pull down more overhead wires before it can be halted.

The possibility of explosions in failed equipment, with the danger to staff and public, makes the failure of electrification equipment more than just the inconvenience of lost train timings. The Railway Inspectorate is often involved in the failures of electrical equipment, either because of injury or because of a dangerous occurrence.

4 WHY DOES ELECTRICAL EQUIPMENT FAIL?

Electrical power equipment consists of copper windings or busbars, paper or polymeric or porcelain main insulation, with oil or high pressure gas or vacuum impregnant, in a sealed, usually steel, tank. The copper busbars or tank offer few reliability problems, but poor seals or corrosion can lead to loss of oil, gas or vacuum, and a failure of the main insulation is then likely.

The main insulation is the weakest link; there are several mechanisms of failure possible:-
4.1- internal discharges
4.2- thermal runaway
4.3- overvoltage puncture
4.4- surface tracking or carbonisation
4.5- surface flashover

4.1 Internal discharges

The electric stress (voltage / thickness) is high in cables and transformers and the constant reversals of voltage (50 Hz) cause small sparks in voids in the insulation. The voids might be present due to poor winding of tapes or impregnation in the factory, due to thermal expansion and contraction, due to electromagnetic forces on short circuit or due to loss of pressure in the impregnant. 33kV cables on Railtrack South Zone have an oil duct in the centre, supplied with high pressure oil, to eliminate voids. Transformers are vacuum treated after winding to remove air voids before oil impregnation. If discharges occur in insulation, they cause damage to the polymeric material, and the void elongates, becoming more severe in discharge and eventually punctures the insulation. Some materials (eg mica) are better able to withstand discharges than others (eg polythene), but insulation is usually designed to be void-free and discharge-free.

4.2 *Thermal runaway*

The constant reversals (50 Hz) of voltage cause movement of ionic molecules in the insulation and the movements manifest themselves as heat and power loss; water ions absorbed into insulation also causes heating. The heating increases with frequency and small high-order harmonics can generate significant heating. The power loss (called dielectric loss or tangent δ) increases with temperature and can become unstable. Electrical insulations (papers, polymerics) are generally low in ionic matter, but oil can quickly pick up water and contaminate the main insulation causing dielectric losses and thermal runaway.

4.3 Overvoltage puncture

High overvoltages occur on electrified lines due to close-proximity lightning strikes or due to switching on inductive (transformer) circuits. On Regional Electricity Boards lightning is the largest cause of outages and equipment failure (2). Fortunately the railway electrification system has high-level earth or booster wires, which act as a screen to induced surges and lightning damage is relatively unknown on railway electrification equipment. However fast-operating vacuum circuit breakers (VCB) can generate high surges on rolling stock without suppression (3). Very high voltages can exceed the electric strength of the insulation and cause puncture. Electrical insulation is usually operated well below is ultimate electric strength, because of void or thermal runaway problems, and overvoltages will most likely cause damage due to initiation of these latter mechanisms.

4.4 Surface Tracking

Most electrical insulation in cables and transformers is polymeric (ie organic / carbon bearing), either paper, wood, epoxy resin or rubber, because these can be wound, formed or extruded into the complex shapes required. Only porcelain and glass used as support insulators are not polymeric. If moisture or conducting pollution settles on a polymeric surface, either by condensation or outdoor exposure, the sparks are energetic enough to cause depolymerisation ie reversion to free carbon. Carbon itself is a conducting material and the sparking progresses slowly in a tree-like pattern called "tracking " taking days to years. Different materials have varying resistance to tracking. In the past only porcelain and glass were suitable for outdoor use, but by proper formulation and evaluation, polymeric materials can now be used outdoors successfully (3). To prevent tracking, it is important to have sufficient distance over the surface between live and earth (the “creepage” distance) for the given environmental conditions.

4.5 Surface Flashover

Tracking can progress across the whole insulator and eventually lead to surface flashover. But if the surface is sufficiently contaminated with salt water or particulate pollution, leakage currrent will flow which produces sparking and immediate flashover, without tracking, even on glass and porcelain insulators. This is a breakdown of the air surrounding the insulator; after the flashover the insulator is usually undamaged and the power can be re-instated quickly.

5 HOW OFTEN DOES ELECTRICAL EQUIPMENT FAIL?

On electrical insulation, surface flashover is the most common failure mechanism, followed by surface tracking, internal discharge, thermal runaway and overvoltage puncture. On the UK rail network of 5000 single track km (STK) of overhead lines and 5000 STK of 3rd rail :-

5.1 About 20 surface flashovers will occur each day on 25kV and 750v insulators due to pollution or debris such as twigs, polythene bags, aluminium cans. No damage is done and the power is re-instated quickly leading to little delay.

5.2 One tracking failure will occur each week on overhead line or 3rd rail polymeric insulators, on switchgear, on transformers or on cable terminations.

5.3 One internal discharge failure will occur each month on paper cables, polymeric cables or switchgear bushings.

5.4 One thermal runaway failure will occur each year due to wet oil in transformers or switchgear.

5.5 One overvoltage puncture will occur each 5 years, probably in switchgear or a transformer, due to lightning or switching surges.

Each of these latter four failures is likely to be explosive, causing damage and extensive delays.

On the overhead current collection system, vandalism is the greatest cause of incidents either by fires, breakages of insulators or thrown debris on the wires; then follows neutral sections, birds, train pantograph problems, insulators and mechanical fittings:-

5.6 One flashover in a tight air clearance will occur every day due to birds; that may lead to wire damage, failure of which then occurs weeks or years later.

5.7 One overhead line dewirement will occur every 2 days somewhere on the system due to all causes.

5.8 One OHL component (dropper, jumper, splice) will fail each week, damaging a pantograph.

5.9 One pantograph will malfunction each month, damaging the overhead line.

The 3rd rail system is more rugged and less liable to damage. After vandalism (fire and theft), most failures arise from clamps and connectors (current rating / heating problems) and train shoegear problems:-

5.10 One shoegear incident each week will cause 3rd rail damage.

5.11 One insulator failure or ramp-end failure will cause shoegear damage each week.

6 HOW CAN FAILURES BE PREVENTED?

6.1 Flashovers

On electrical insulators, surface flashovers can be prevented by the use of long creepage. Creepage on porcelain insulators was reduced in 1970's from 1200mm to 790mm, leading to simpler, smaller and cheaper insulators. Using polymeric insulators, creepage can be increased for a given insulator length by the use of thin and wide sheds, which would not be viable in porcelain. Where flashovers occur in heavy salt areas (seaside, industrial) doubling of insulator length or the use of thin, long-rod insulators is needed.

6.2 Tracking

Surface tracking can be prevented by the use of the correct material for the environment. Where any condensation on surfaces can occur (outdoor switchgear), phenolic laminates are not adequate and track easily; glass-based epoxy laminates are required. Where surfaces can be wetted by rainfall, only the best polymeric materials, such as silicone rubber or ptfe, can be used. Epdm's, epoxies, polyurethanes, polythenes are not suitable and will track or erode. In oil-filled equipment such as transformers, which breath often due to thermal cycling, excessive water in the oil can lead to tracking of paper laminates. Adequately-sized air driers and regular maintenance will prevent wet oil.

6.3 Discharge

Internal discharge can be prevented by correct design and manufacture of bushings, cables and transformers. In oil-pressure cables it is important that the pressure is maintained else voids will appear and discharges will begin. In transformers it is important that oil levels are maintained, so that the windings are not exposed to air and voids produced; transformers must be refilled quickly after draining to prevent voids forming. Standard tests are available to evaluate internal discharge severity. If discharges occur in oil-filled equipment, oil analysis can be carried out for discharge products.

6.4 Thermal runaway

Thermal runaway can be prevented by keeping water out of insulation. Most water enters via the oil, which collects water through inefficient breathers. Oil quality is most important. If the oil is wet, thermal runaway **and** tracking can occur. Water in oil causes accelerated ageing of paper insulation ie embrittlement and the paper is not able to withstand winding movements, due to electromagnetic forces, and punctures. Standard tests are available to measure oil quality and degradation products. Oil analysis can also be used to detect paper deterioration.

6.5 Overvoltages

Overvoltage puncture can be prevented by suppression of surge voltages by spark gaps or surge arrestors connected to the overhead line or on vehicles.

6.6 Vandalism

On overhead lines, vandalism can be reduced by removal of debris from trackside used for throwing and by fencing to restrict access. Electrocution incidents can be reduced by reducing the amount of live metalwork, such as by using earthed cantilevers. Porcelain insulators are particularly vulnerable to vandalism; breakage of sheds can lead to core fracture, with the cantilever frame collapsing onto the wires and damaging the pantograph. The rubber shed protector (4) was developed to prevent damage to porcelains and has been used widely in urban areas with great success. Polymeric insulators are also vandal-proof and have been used in electrification schemes in Birmingham and Leeds.

Both vandal debris and birds can lead to flashovers at overbridges; the arc can damage strands in the catenary wire, which may fail months later. The use of solid wires, not stranded, is recommended at bridges. Birds can also cause flashovers at tight clearances on insulators and pulleys on Mk 1 portals; the arc can damage strands at the wire / pulley wheel interface, which then fail much later due to fatigue.

6.7 Neutral sections

Neutral section glass-fibre insulators have a high maintenance requirement and a high failure rate; failures can be minimised by regular cleaning of the surface, correct positioning of the arc-horns to prevent thermal shock damage when flashover occurs, regular checking of track magnets to ensure no arcs are drawn and regular attention to wire profiling to avoid impacts on the insulator beads.

6.8 Droppers and jumpers

Failure of droppers and jumpers can be catastrophic, since these wrap around the pantograph and cause a dewirement. Regular inspection for looseness and fatigue is required. Bolted current-carrying clamps can loosen or corrode with age, leading to overheating and fracture; most bolted clamps on the bi-metal Al / Cu Mk 3 system were replaced soon after installation by C-spring wedges, which do not deteriorate.

6.9 Wear

Wear of the overhead wire is very slow (life > 50 yr on average) and failure of the wire due to fracture is unlikely; however at hardspots, such as registration arms, crossovers and bridge supports, wear is increased and "necking" failures do occur due to poor profiling of the wire.

6.10 Pantographs

Badly-performing pantographs cause dewirements, due to increased upthrust from badly-set aerofoils, pressure regulators or springs. These pantographs run adequately most days, but in adverse conditions such as high winds, overspeeding or coupled with poor wire profiling or tight clearances, they can foul structure metalwork, leading to pantograph damage and consequently dewirement.

6.11 Wind

Wind or badly-aligned wire relative to track can cause the wire to slip off the pantograph head, allowing the pantograph to rise above the wire and hit the next structure. Faults due to wind can be minimised by correct OHL design based on a knowledge of wind speeds and topography throughout the route and by wind monitoring at exposed sites to relay extreme conditions to Control Centres to slow or halt trains to prevent dewirement. Failures due to poor wire alignment can be minimised by regular checking of the OHL wire height and stagger, either by gauging trains or better by service trains with measuring pantographs.

6.12 3rd rail

Third rail systems are more rugged and will tolerate damaged shoe-gear; indeed it is the shoe-gear which suffers in the event of poor current collection rather than the current feeder system as on OHL. Poor current collection can be due to movement of the 3rd rail relative to track, either vertically or horizontally, usually due to several damaged insulators in a row or due to debris on the rail, either vandalism or blown. Prevention of faults should be by regular measurement by gauging trains or better by service trains with measuring shoe-gear.

7 DIAGNOSTIC TESTING

Equipment reliability can be improved by diagnostic testing, also called condition monitoring All electrical equipment has a maintenance period recommended by the manufacturer, based on experience with similar equipment in a similar working environment. It is the responsibility of the Maintenance Engineer to ensure that the maintenance is carried out and that this is the correct maintenance period, based on failures in service, and increase or decrease the period as necessary. Also additional more-sophisticated testing of equipment can be undertaken to identify weak or abnormal units. But the testing must be relevant to the eventual failure mechanism; this means that failures must be investigated and the cause established. If the failure is due to vandalism rather than natural ageing, then diagnostic testing the remaining equipment will not prevent failures! Good record keeping of incidents, failure investigations and diagnostic test results is essential if improvements are to be made to equipment via focussed research.

7.1 Electrical insulation

Insulator flashovers can only be predicted by using local weather and pollution monitoring, relayed back to Control Centres, but they cannot be prevented. Prevention is by the correct choice of insulator creepage for the local conditions at the design stage (see IEC 815), or in emergencies, by greasing to prevent the pollution wetting. Standard tests (eg IEC 507) are available to evaluate designs for different pollution severities.

Surface tracking can only be prevented by the correct choice of material for the pollution conditions. Tracking of materials in wet oil in switchgear and transformers can be prevented by regular oil sampling and testing to IEC 422 for electric strength and water. Monitoring of oil for wetness will also indicate whether thermal runaway of insulation is likely; wet oil can and should be dried by regular recycling through vacuum drying plant. Standard tests (eg IEC 587) are available to evaluate materials for particular applications.

Oil sampling with dissolved gas analysis (DGA) to IEC 567 can indicate internal discharges or surface tracking or overheating at poor metal contacts. The analysis can be done on oil samples sent to a laboratory, from power transformers, boosters or switchgear, or larger and more important transformers can have electronic DGA monitors fitted to the oil system for continuous recording and remote data transfer. Oil analysis can also be used to detect furfuraldehyde (FFA), a by-product of paper deterioration, which can be used to identify weakened transformers (5). Most UK railway transformers have been tested for DGA, electric strength and water regularly.

Internal discharges in bushings and cables can be detected by discharge detectors. These might be ultra-sonic devices or Transient Electromagnetic Voltage (TEV) radiation detectors used on the earthed screens of equipment or high voltage tests using tuned detectors as in IEC 270. The latter gives absolute measurements, whereas the former give relative measurements, enabling a ranking order for attention to be determined. All have been used widely on the UK railway (6), particularly on WCML oil / paper switchgear in the 1980's and on 33kV Railtrack South Zone switchgear. They can identify the insulation with the severest discharges, which can then be removed from service before failure.

Thermal runaway can be detected by dc leakage current testing or ac dielectric loss / tan δ testing. These need specialist high voltage testing, such as was used on WCML bushings (6).

Overheated connections in switchgear busbars, transformers or isolators or in overhead line or 3rd rail connectors can be detected in service by thermal vision cameras if the joints are exposed or by oil analysis if the joints are in tanks. On de-energised equipment, high joint resistances can be measured by micro-ohmmeters or "ductors".

7.2 Current Collection

On OHL systems, poor pantograph performance can be detected by equipments such as "Panchex" (7), which monitors the uplift of the contact wire as a pantograph passes. Other interactive parameters could be measured, such as wire oscillations or acceleration, bad performance being indicated by a result well-removed from the norm or mean for that type of pantograph at that speed or for all pantographs.

Poor overhead line condition or hard spots can be identified by equipments such as "Olive" or gauging trains such as "Mentor" (8), which measure impacts by an accelerometer on the pantograph head; "Olive" is mounted on service trains and gives a daily statement of the condition of the OHL, and via GPS (global positioning system) gives the location of hard-spots and possible incidents.

Similar shoegear and 3rd rail monitoring equipment could be developed to improve 3rd rail current collection performance.

Automatic monitoring of the mechanical condition or imminent failure of other overhead line or 3rd rail components is more difficult to realise.

8 DESIGN IMPROVEMENTS FOR THE FUTURE

Experience has shown where weaknesses exist in electrification equipment. These weaknesses should be designed out in the next generation of equipments.

8.1 Insulators - polymeric insulators should be used throughout to prevent breakage due to vandalism; they should be incorporated in novel arrangements such as earthed cantilevers, to prevent breakages due to snatch bending loads. The earthed cantilever has less live equipment and will reduce flashovers due to blown debris and prevent many touch incidents due to vandalism. Debris on trackside for vandals to throw should be cleared regularly.

8.2 Wires - solid copper wires should be used widely on catenary to prevent breaking of strands due to bird flashovers at tight bridges and flashovers and fatigue at pulleys. Pulleys with their seized pivots or need for regular greasing should be eliminated. Wire height and stagger and impacts should be monitored regularly by instrumented service trains to anticipate faults. Mechanical independence for overhead wires on each track is essential to prevent dewirements halting all electric trains.

8.3 Droppers - a flexible dropper, clamped at both ends to act as dropper and continuity jumper, should be used to eliminate dropper breakages and uncouplings.

8.4 Neutral Sections - should be replaced by overlaps or a much-improved insulator developed.

8.5 Wind Problems - pantographs should be used in their optimum aerodynamic direction only, using two pantographs per motive power unit. Pantographs should have a more sensitive dropping system, operating automatically if poor current collection (ie excessive arcing or excessive head acceleration) is detected. Regular pantograph monitoring in service is recommended, using automatic on-line techniques. Wind speed recording at key locations is necessary to enable decisions to be made by Control Centres as to whether to slow or halt trains in high winds. Fundamental work should be undertaken to examine the movement of wires in wind and the mechanism of dewirement. Wind blown debris should be removed from trackside regularly.

8.6 Substations - highly-stressed insulation, such as cables or bushings, should be minimised in substation design. Oil-insulated equipment, with its inherent explosion hazard and regular maintenance, should be replaced where possible by vacuum switchgear and dry-insulated systems. Where oil is used in transformers and switchgear, its quality is of paramount importance and regular sampling and testing for water, electric strength, DGA and FFA should be carried out to anticipate and identify failures.

8.7 Fault Location - the cause of more than half of all circuit breaker trippings is unknown, but each fault represents possible damage to equipment. Accurate fault location devices should be installed and all circuit breaker trips should be located and investigated on-site to look for possible damage and so anticipate failures.

8.8 Alternative routes - because loss of power supply or overhead line dewirement can halt **all** electric trains, alternative routes and links must be electrified to maintain the service.

9 CONCLUSIONS

The causes of failure of existing UK railway electrification equipment have been described and the techniques available to monitor its condition. Since the life of electrification equipment is longer than the working life of an engineer responsible for it, good record keeping of design, testing and failure data is essential to pass on the experience gained. Detailed examination of failures to establish their cause is also essential to ensure the correct maintenance decisions are taken. The paper has identified areas where design improvements can be made on new electrification schemes and where research needs to be carried out to improve understanding.

REFERENCES

1 Atkinson W.C. and Ellis F.E.
"Electricity Distribution - asset replacement considerations."
IEE Electronics and Power, May 1987, pp 337-340.

2 IEE Conference on "Lightning and Electricity Board Performance"
London 1984.

3 Bradwell A. and Wheeler J.C.G.
"Developments in glass-fibre bridge and tunnel insulators for minimum clearances on British Railways."
Proc. IEE., Vol. 131, Pt. B, No.1, July 1984, pp 1-6.

4 Wheeler J.C.G.
"Anti-vandal shed protectors for solid-core porcelain insulators on BR."
IEE DMMA Conference, Lancaster, UK. 1984.

5 Shroff D.H. and Stannett A.W.
"A Review of paper ageing in power transformers "
IEE Proc., Vol. 132, Pt. C, No. 6, November 1985.

6 Bradwell A. and Bates G.A.
"Analysis of dielectric measurements on switchgear bushings in BR 25kV electrification switching stations."
IEE Proc., Vol. 132, Pt. B, No. 1, January 1985, pp1-17.

7 Bradwell A.
"Monitoring the overhead - uses of Mentor, Panchex and Olive."
ICE Seminar on "Measuring, monitoring and recording on track."
London, UK. 2 July 1996, Paper 3.

8 Betts A.I., Bates G.A., Hall J. H. and Keen P.M.
"Pantograph and Overhead line Monitoring."
IMechE Seminar No. S079, "4 September 1992, London, UK

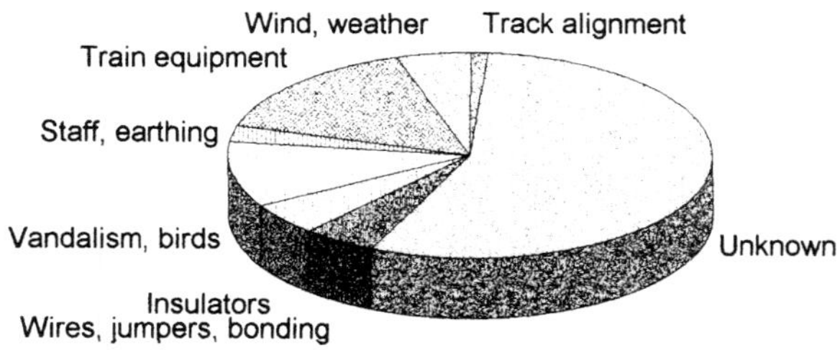

Figure 1 Causes of OHL Incidents

C570/017/99

Control of earth leakage currents and track potentials on a.c. and d.c. electrified railways for a fault-free infrastructure

R D WHITE
WS Atkins Rail, Derby, UK
D C KNIGHTS
WS Atkins Rail, London, UK

ABSTRACT

A.C. and d.c. railway systems are unique amongst electrical installations in that the return path intentionally includes the earth in a.c. systems, and unintentionally in d.c. systems. Since the railway is an integrated system it is fundamental that the system and the associated structures must be earthed to the ensure safety of the railway and to reduce the levels of touch and accessible potentials.

The control of the earth leakage currents and track potentials are essential if a fault free electrification infrastructure is to be provided. Earth currents and adverse rail potentials can be responsible for the mal-operation of line-side equipment, the interference into safety critical circuits and, on d.c. systems, for the corrosion to railway and third party equipment. The earthing arrangement on d.c. systems is compromise between minimising the level of rail to earth potentials and eliminating the corrosive effects on earthed structures.

1 LEGISLATION: EUROPEAN AND RAILTRACK STANDARDS

The earthing and bonding of the railway system are designed in accordance with the relevant European Standards, National Standards and regulations. Please refer to References for a summary of European Legislation, UK and Railtrack Standards.

2 EARTHING OF 25 kV 50 Hz ELECTRIFED RAILWAYS [Figure 1] (8)

The 25 kV railway overhead system is a single pole distribution network with the neutral at the feeder station earthed. Traction earth is defined as 'the distributed earth system formed by the traction return rails intentionally connected to the general mass of earth by the foundations of the overhead line structures' Reference (1).

On single rail return systems, one of the running rails of each track is designated as the 'traction current return rail'. These rails are all bonded together and to the adjacent overhead line structures. A cross bond connects one or both running rails of one track with one or both rails of another track at intervals of about 500 m equally spaced between the booster transformers. The structures of the railway system are also bonded to the return running rail on single rail return system. On double rail return system [areas of double rail track circuits] the traction return continuity is maintained by the use of impedance bonds at the end of each track circuit. Earth wires are used to connect the steelwork of two or more overhead line structures and bonded to the return rail via an impedance bond. The bonding of the electrification structures is required for safety and to minimise the touch and accessible potentials.

Impedance bonds are used on tracks where both the rails are traction return rails and are equipped with double rail track circuits, all bonding to the traction return rails shall be made via an impedance bond. The impedance bond is used to separate the traction return current and the track circuit current.

A bond required by the signalling system for the integrity of the track circuits is designated as a 'yellow bond'. A bond that if disconnected could result in the bond itself or the associated equipment rising to a dangerously high potential is designated as a 'Red Bond'. (1)

Since the foundations of the structures have a depth of 2 to 3 metres, weather conditions have little effect upon the resistance of a typical foundation (approximately 20 ohms). The total earthing system is a distributed network with a resistance to earth of less than onc ohm. This resistance is of a low enough value that it is not necessary to drive additional earth rods.

The main factor to be considered with an earthed electrification system is the potential of exposed metal in relation to the general mass of the earth. Under normal operating conditions this should be limited to 25 volts (under the new European Directive EN 50122 (5), (6) this has been increased to 60V), however, under fault conditions the limit rises to 430 volts. 'Accessible voltage' is defined as the rail potential under operating conditions that can be bridged by a person. The 'touch potential' is the same situation occurring under fault conditions.

3 EARTHING OF D.C. ELECTRIFIED RAILWAYS [Figure 2] Reference (3,4,7,8)

With high-density urban d.c. electrification systems there is no unique earthing arrangement applicable for all electrification schemes. Typical arrangements to control earth leakage currents and track potentials include: fourth rail systems, d.c. supply negative is earthed at the substation, totally floating earth, earth connected directly to the substation negative drainage diode earth, floating earth with rail potential control devices, and stray current collection system.

A third rail d.c. electrified railway is intentionally earthed permanently at the substation, this is necessary to minimise the leakage of electrical charge and the consequential damage that may be caused by corrosion, and to control the touch and accessible voltages. The railway system returns the traction current to the substation via one or both of the running rails with leakage to the earth via the track and sleepers [typical leakage resistance 8 ohms/km]. To minimise the stray d.c. current the railway infrastructure including overhead line masts, station platforms, signal gantries, etc. are earthed to a separate earthed conductor. With d.c. electrified railways it is essential that the rail touch and accessible voltages are minimised, and at the same time control the leakage of d.c. return currents from the return conductor running rails.

The railway normally uses the train wheels and running rails as the traction current return path to the d.c. substation, (except fourth rail systems). The rails are not normally insulated from the sleepers or structures on which they are fixed. The earth return is considerably more complex than is at first perceived since the current that flows out of the rails will return at some other point, and this is referred to as stray current.

There are typically five earthing strategies associated with d.c. electrified railway systems:

Fourth rail systems:

- The problem relating to leakage currents is overcome by the installation of a the fourth rail (London Underground UK)

Earth system using the Running Rails as the return conductor:

- Totally floating Earth (Sheffield Super Tram UK),
- Earth connected directly to the substation negative (Railtrack UK),
- Drainage Diode earth Docklands Light Rail, Croydon,
- Floating earth with Rail Potential Control Devices,
- Floating Negative Automatic Ground Switches (Ankara Rapid Transit System),
- Over Voltage Protection Device (Lantau and Airport Railways Hong Kong),

If the track is insulated from ground to minimise leakage, the presence of insulation means that voltages between the track and ground will occur. These voltages are a potential hazard to passengers and railway staff when coming into contact with rail or anything connected to it and the ground. Reduction of rail touch voltages can be achieved by clamping the rail to ground if a dangerous voltage is reached and resetting as soon as possible afterwards. It is necessary to have quick detection of the high voltage, which then activates a "clamp", by a thyristor device, GTO device or contactor, to short out the voltage.

Earth leakage computer studies have indicated that the totally floating earth is the most effective in terms of stray current control, with the directly connected earth the scheme produced the highest gross leakage charge. With the Rail Potential Control Devices this proved to be an effective scheme for rail potential control. Rail Potential Control Devices have the ability to make the system totally floating by setting the control voltage at a high threshold.

3.1 Stray current and corrosion [Figure 3]

The railway authority is responsible for any damage or corrosion to other companies' equipment due to the effects of stray or leakage currents from the railway. With new railway electrification projects this has to be addressed during the design period, to minimise any modifications that may be required in operation of the railway.

In many systems the negative of the d.c. supply and the running rails are connected to the earth. Thus the earth may provide an additional path, in parallel with the track, for current flowing from the traction unit back to the point of supply. In cases of metallic pipe lines or cables having metallic armour of sheaths laid in the earth, part of the current may be picked up at some point on the conductor and discharged at another, and this will lead to accelerated corrosion at the point of discharge.

3.2 The stray current collection system (SCCS) [Figure 4]

SCCS can be constructed in the bed of the rail track. It is achieved by the inclusion of steel reinforcing bars in the track bed forming a collector mat located in the track bed below the rails. This mat is bonded together to form a continuous electrical conductor for intercepting the stray current and providing a path back to the substation.

Typical arrangement, [Docklands Light Rail] of the longitudinal reinforcing bars is that they should be not more than 150mm apart. The traversing bars at intervals of typically 100m each being connected to the longitudinal bar and then to a copper bar which is taken to the surface of the concrete, where it is connected at a suitable point to the Stray Current Collector. This is a conductor, which runs continuously throughout the system providing a parallel connection.

3.3 Direct Electric Drainage:

Corrosion can be reduced by bonding of track-side equipment such as pipelines cable sheaths to the return rail at the more negative portion of the track; ie near substations or where negative feeders are connected to rails. The bonding will drain some of the leakage current back into the rail and thence back to the supply thereby ensuring that the structure (that is bonded) receives partial or sometimes complete cathodic protection.

Direct electric drainage cannot be used in all circumstances where there is a likelihood of current reversal in the drainage bond. A rectifier or diode is usually provided for this purpose. It may be necessary to limit the current by the inclusion of a series choke, resistance and fuse with the diode. Similar diodes are connected to the substation earth to provide a return path for the system d.c. earth faults. Drainage diodes are rated for the short circuit level for a fault between the traction supply positive and the Stray Current Collection or earth.

4 EARTHING AND CONTROL OF STRAY CURRENTS

The earthing and bonding system is required to ensure a safe railway and fulfil the following functions (1):

- To provide adequate continuous return circuit for the return of a.c. traction load current to the feeder station. [a.c. electrification system only].
- Non overhead line structures required to be bonded [a.c. electrification]. To avoid the risk of electric shock and to provide a path for fault current in the event of a flashover from the a.c. overhead line or due to contact with a broken contact wire: Metal bridges or pipes spanning the overhead line (gas, water, sewage, telecommunications, electricity), signal gantries, station structures, steel or reinforced concrete structure over or underneath overhead line, metal pipes fences or cables along the track-side and within touching distance the overhead line masts, electrical distribution wiring on railway premises.
- To provide an adequate connections to the remote earth by way of the foundations of the overhead line masts.
- To ensure that there is adequately rated return path for fault current that may arise from a flashover or broken conductor or de-wired pantograph.
- To ensure that the touch voltages do not exceed those specified for safe operation under normal and fault conditions.
- To ensure that potentials arising from capacitive effect do not exceed safe operating level.

- To ensure that return currents and fault currents do not return to the feeder station via non specified route i.e. metallic services, fences etc.
- To ensure the safe operation of the electrical supply, in the vicinity of flammable gases liquids and oxygen etc
- To ensure correct interfacing between electrified railway networks, and that there is safe operation of all railway track circuits within the vicinity of the railway a.c. /d.c. interface.
- To minimise interference into the railway and public telecommunications network,
- Control the return of d.c. traction current to minimise the effect of stray return current, capable of producing corrosion (d.c. electrification system only).

Some railway authorities have overcome the need to provide an earth return by the introduction of a fourth rail (LUL), or an alternative arrangement is to insulate the track and run a power system return which is earth free.

5 EARTHING AT A.C./D.C. ELECTRIFICATION INTERFACE

It is necessary in some locations to operate a.c. overhead lines and d.c. conductor rails on the same track or adjacent tracks. The most common reason is to allow dual voltage train to change over from 750Vdc to 25kV 50Hz overhead line. It may be necessary to bond the rails of the two systems together to avoid excessive touch and accessible potentials.

At an a.c./d.c. interface the return current from the a.c. and d.c. traction units may be required to flow in the same rails. This is due to the requirement for there to be a path for current to flow to the a.c. feeder station and to the d.c. sub-station. It is inevitable that there will be a path for the a.c. return current to flow into the d.c. only area and for the d.c. return Current to flow into the a.c. only area.

This operation is hazardous as the a.c. current flowing in the d.c. only area may cause disturbance to the 50 Hz track circuits. D.C return current in the a.c. only area is undesirable since the stray d.c. current flowing in the overhead line structures and other bonded equipment will produce electrolytic corrosion where the current flows out of the metalwork.

The stray current at an a.c/d.c. interface can be controlled by the specific design of the two electrification systems at the interface. The following methods can be used and may be implemented singularly or in combinations.

- The use of insulated rail joints in both the positive conductor and return conductor rails to prevent the flow of direct current.
- The use of rail joints bridged by resistors or other impedance to limit the flow of d.c. current to an acceptable level.
- The segregation of structural steel work and earth's of electrical systems from the rails, thereby limiting the flow of d.c. current through such items.
- Suppression of stray current by active control, using voltage-controlled substation.
- The use of isolating transformer fed-sections for high voltage a.c. electrification system.
- The use of filters to isolate the d.c. from the 50 Hz electrified railway system.

On tracks equipped with both a.c. overhead and d.c. conductor rails, the design for the earthing and bonding shall be to provide safety to persons in the presence of the 25kV electrification system in both normal and short circuit conditions. The bonding shall ensure rapid clearance of short circuits, and not provide a path for d.c. return current to flow in the railway or third party metallic structures. The bonding should ensure that there is no significant damage by stray current to line side structures.

The bonding in a.c. /d.c. areas may include the omission of bonds where they would normally be provided. The a.c. / d.c. electrification may be required to operate without the benefits of bonding to structures and electrical distribution earths. The Earthing and Bonding plans shall ensure that the traction return currents in the rails, including effects relating to broken rails and disconnected continuity bonds and cross bonds do not result in accessible voltages which exceed limits specified in Appendix B RT/ES/21085.

Voltage can be limited under short circuit conditions using voltage-limiting devices such as spark gaps and non-linear resistors to ensure compliance with limits stated in relevant earthing standards. References (1, 2, 6).

6 ENGINEERING DESIGN FOR FAULT FREE INFRASTRUCTURE, SYSTEM DESIGN

Modelling of railway systems is required to investigate potential failure modes, and predict the effect of failure on various aspects of the electrification system. Modelling techniques are available to predict level of earth leakage currents, the interference of a.c./d.c. leakage currents with signalling track circuits, the levels of accessible and touch potentials, and the interference into telecommunications and signalling systems. A 'fault free' electrification infrastructure is vital if the operation of the railway is to be safe, reliable, available and maintainable.

6.1 The engineering design process

Design engineers are required to specify the operating characteristic and performance of electrical systems prior to installation or modification. They must ensure that the strict requirements of the railway and European EMC standards are adhered to. Modelling of railway systems is able to provide supporting evidence of the operating characteristics at the design stage of the project and hence minimise the costs that may be incurred by modification to systems after commissioning.

An example of the modelling for a 25kV 50 Hz electrification system

The simulation of electrification systems includes the modelling of the overhead line equipment, conductor and track impedance, and mutual inductance between conductors, booster transformer, return current system and the bonding arrangement. The characteristic value of the circuit impedance/km is evaluated using the known equations for separation of conductors and the Carson Pollaczek formulae. The specific value for each element of the along track impedance is calculated by this method. The impedance matrix is assembled by including the self and mutual values for the discrete lengths for each section of the electrification OLE and return conductor system.

The connection of the feeders, booster transformers, cross bonding, earth wire connections must be specified. Once the impedance of all the circuit elements and interconnections are known the model is connected. The nodal admittance matrix is assembled in terms of the series and parallel circuits representing the electrification system. The boundaries of the model and leakage resistances to earth are required to be added to the nodal admittance matrix. The positions of loads or short circuits are included and mathematical matrix routines are then used to solve the complex matrix equations. The results from simulations will contribute supporting evidence in the preparation of safety requirement specification and acceptance plans, for the introduction of the system being simulated.

Requirements for the railway business
E.M.C. simulations will give confidence to the operational side of the business and minimise the risk associated with E.M.C.. Careful hazard analysis, simulation studies and design of the system will provide performance characteristics and mitigate, against incompatibility. The typical railway business requirements include:

- Information required in the preparation of a 'safety case'.
- Due diligence and compliance with E.U./B.S. standards.
- Minimise the technical risk of malfunction of the equipment,
- Minimise the risk of failure due to incompatibility.
- Minimise the commercial risk of the failure of equipment owned by a third party.
- Minimise the commercial risk and loss of revenue due to the failure of equipment.
- Minimise the risk of failure or interruption of safety critical circuits.

6.2 Hardware analogue modelling tools (7)

Hardware modelling of a railway infrastructure will enable the investigation of the system on a 'reduced scale'. The model can identify characteristics that the software model is normally unable to observe, including commutation failure in inverters/rectifiers or instability within the control system etc. The hardware model is able to predict accurately the behaviour of the power system, load and control system, with the circuit being required to perform precisely the same function on a reduced scale. If all the circuit parameters of the scaled model are scaled correctly with the inclusion of the control system then the model should reflect the real system. The detail in the modelling process is fundamental to ensure accurate results are obtained from the model.

The model is required to operate at a reduced voltage and current. Careful design is required to ensure that the scaling of the component values is correct, the hardware can be assembled quickly, from reusable components. The component values of parameters in the model can be adjusted relatively easily. Re-scaling the model voltage, current ratios or remodelling from a 10 to 20 MW system would require a redesign of the complete system. The results from the hardware model will normally be in real time and provide the complete response, i.e. the transient and steady state response for the network.

6.3 Computer based 'software' modelling tools (7)

Software modelling tools includes written computer code, the application of bespoke packages, circuit analysis packages, and mathematical packages. Computer languages are the most basic of the tools requiring written libraries that perform numerical and integration routines. These languages are the building blocks for the more advanced 'software packages'.

Mathematical packages have been developed with mathematical routines embedded within the software. The mathematical package is simplified by having an input procedure that is similar to writing mathematics on a piece of paper.

The circuit analysis packages are the most powerful tool available and the most expensive. The main advantage of the circuit analysis package is the simplicity of inputting the circuit in schematic form. The package also has all of the numerical analysis routines embedded and some have the ability to combine electrical and mechanical simulations.

6.4 Typical simulations including failure scenarios on a.c./ d.c. electrification systems

Example of typical calculations that are required in the design process or modification of an electrified railway system include:

- Short circuit and load calculations.
- The level of conductive and induced rail to earth & voltages into line-side equipment
- The control of earth currents at a.c. /d.c. electrification interface.
- The control of 50 Hz traction return currents into 50 Hz a.c. track circuits.
- Interference into d.c. track circuits [a.c. electrified railways] by d.c. traction current.
- Electro-magnetically and conductively induced voltages into signalling circuits.
- Two-dimensional modelling of the tracks and overhead lines.

7.0 CONCLUSIONS

The control of the earth leakage currents and track potentials is essential if a fault free electrification infrastructure is to be provided. Earth leakage currents and track potentials can be responsible for the mal-operation of line-side equipment, the interference into safety critical circuits and the corrosion to third party equipment. Computer modelling is part of the overall design process and can be used to minimise the risk associated with the control of earth leakage currents and excessive track potentials.

8.0 REFERENCES

UK/EU Standards and UK Regulations on earthing include the following specifications:

(1) RT/E/S/221085 Design of Earthing & Bonding Systems for 25kV a.c. Electrified Lines
(2) RT/E/S/21032 Earthing Systems for d.c. Substations
(3) RT/E/S/21104 Design & Installation of Electric Track Equipment d.c. Electrified Lines.
(4) RT/E/C/27022 Design and Installation of negative bonding on d.c. electrified lines.
(5) GM/RT/1010 Electrified Line Traction Bonding.
(6) EN 50122 Part1, European Directive on Earthing and Bonding.
(7) 'Modelling Techniques for EMC Studies on High Performance Railway Systems' Dr R D White; Electrification & Power Supply AIC Conference 1999 London.
(8) 'D.C. Railway Electrification and Protection' Dr R D White; IEE Traction School '99.
(9) 'A.C. Railway Electrification and Protection' Dr R D White; IEE Traction School '99.
(9) Construction Regulations (Design and Management) 1994.
(10) Railways (Safety Case) Regulations 1994.
(11) Institution of Electrical Engineers' Regulations.
(12) Electricity Council Engineering S 5/1.
(13) BS 7430:1991 Code of practice for earthing.

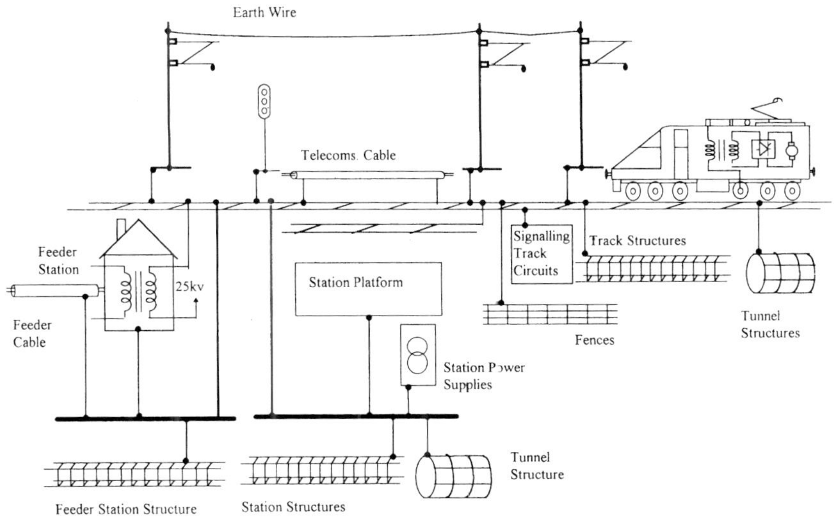

FIGURE 1 EARTHING AND BONDING IN AC ELECTRIFIED RAILWAYS

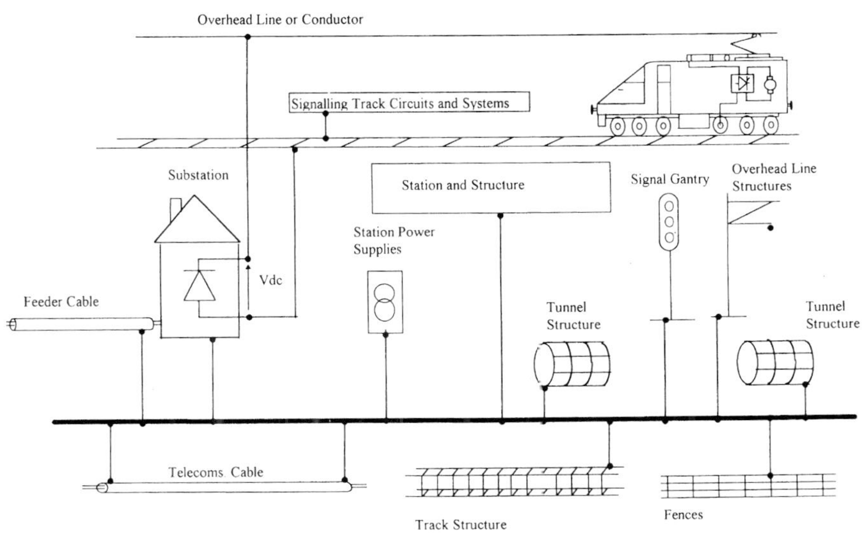

FIGURE 2 EARTHING AND BONDING ON DC ELECTRIFIED RAILWAYS

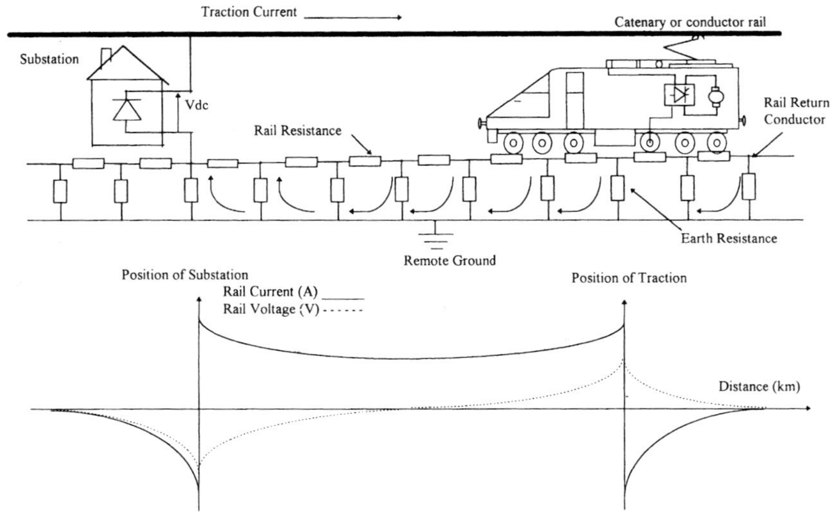

FIGURE 3 EARTH CURRENTS IN D.C. ELECTRIFICATION SYSTEMS

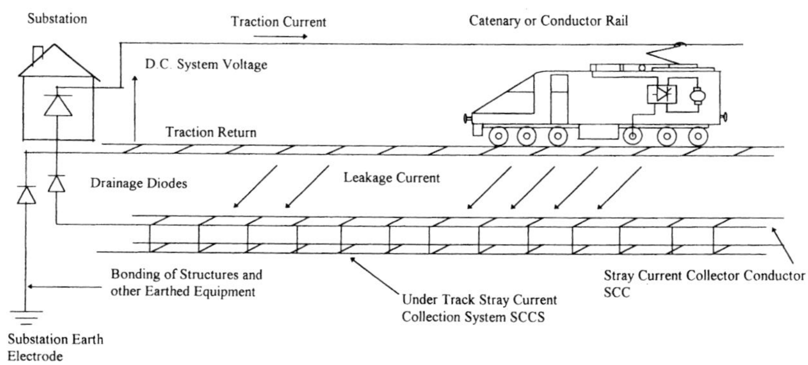

FIGURE 4 TYPICAL EARTHING SCHEMATIC INCLUDING DRAINAGE DIODES AND STRAY CURRENT COLLECTION SYSTEM

C570/019/99

Improved reliability of escalators in the London Underground infrastructure

S MAW
Neale Consulting Engineers Limited, Farnham, UK
G IZIENICKI
London Underground Limited, London, UK

SYNOPSIS

London Underground Limited (LUL) are responsible for the operation of approximately 300 escalators within their system, the reliability of which is essential for efficient passenger movement within stations.

LUL in conjunction with Neale Consulting Engineers Ltd (NCEL) initiated a program to identify the major causes of downtime and this paper presents the investigation process involved to identify the key problem areas within the escalator machinery and where improvements were possible.

The investigation undertaken by NCEL demonstrated that a prime cause of downtime was unpredictable wear behaviour of the large step chains which move the steps around the machine. Examination and measurements of a large number of replaced components identified the mode(s) of wear and further analysis identified that improved wear resistance, and more importantly, consistent wear performance could be achieved by the use of polymeric based materials.

Test rig work demonstrated that an in-service life in excess of 40 years was achievable and full size chains were therefore installed into two working escalators. These polymer bush chains have now been in service for a number of years and the monitored wear performance confirms that escalator chain performance is anticipated to be in excess of 40 years.

The success of the test programme has lead other Metro systems and also escalator manufacturers to follow suit and polymer bush step chains are now also in service in New York, Paris and various other cities. The long term effects of this innovation are expected to include substantially reduced downtime and decreased total life costs.

INTRODUCTION

The rapid movement of large numbers of passengers on the deep level lines of the London Underground system, requires the extensive use of escalators to achieve the required passenger flow rates, between platform level and the surface. London Underground Limited (LUL) are responsible for the operation of approximately 300 escalators, with another 107 machines joining the fleet upon the opening of the Jubilee Line extension. These are an essential and integral part of their system, and escalator reliability is one of the determining factors in the efficient operation of the whole of this transport system.

In a situation of this kind, it is essential to have an effective policy for escalator maintenance and replacement, guided by an understanding of the operation and failure mechanisms of their critical components, so that adequate resources can be directed to keeping these components, and therefore the escalators, in a fully serviceable condition.

Neale Consulting Engineers Ltd. (NCEL) have been working closely with London Underground Ltd. for approximately 10 years to improve the reliability of the escalators in the LUL system, by feeding in the extensive international experience of NCEL in the failure and design improvement of all kinds of mechanical plant. This paper describes the methods used in this joint investigation and the results obtained, with their potential for giving major improvements in system reliability.

RELIABILITY ANALYSIS

As a starting point for the study, London Underground made available data on all their escalators giving the location, rise height and design type. They also gave access to all their maintenance records so that the causes of unreliability could be analysed. Neale Consulting Engineers used all this data as a starting point and also visited twelve selected underground stations to observe the machinery in operation and under maintenance. This provided an opportunity to understand the detailed operating environment as well as the maintenance techniques, and to see worn components immediately after their removal.

London Underground escalators, such as the type shown in Figure 1, are of a more robust design than those commonly used in shopping centres and department stores and have double passenger width. Rise heights are up to 27 metres, and some of the current machines were originally installed in the 1920s. The robustness of the machines is indicated by the fact that the average operating life of a good escalator design is of the order of 40 years, before it requires complete overhaul or replacement.

The operating records for all London Underground escalators were examined to check the level of reliability being achieved. This showed that the total number of escalator operating days lost per year was 3050 which corresponds to an average of about 11 days lost per escalator per year.

The maintenance records were therefore analysed in detail to determine the individual components that were causing the greatest total out of service time in an operating year. This showed that the greatest problem arose with steps, chains, combs and skirts which accounted for an average out of service time of about 8.5 days per year per escalator. This was therefore analysed further to determine the detailed nature of the various occurrences. This showed that of the 8.5 days, about one third arose from damage to combs, treads, skirt guards and panels. This kind of damage tends to be passenger related or arises from the use of the escalators during the night to carry cleaning and servicing equipment in and out of the stations.

The major engineering related failure which also accounted for a third of the out of service time were problems related to the step chains. Each escalator has two of these roller chains with several hundred links, bushes and pins, and these chains drive the steps around the escalator. The remaining third of the out of service time, relating to the step system, arose from problems with all the other components such as steps, step wheels and their tracks, and the chain sprockets and their support systems at the end of the escalator.

The reliability analysis indicated therefore that the focus for an engineering improvement for greater reliability should initially be focused on the step chains.

A STUDY OF STEP CHAIN PROBLEMS

Escalator step chains in common with other roller chains, as used on bicycles and other chain drives, tend to wear at their pins and bushes resulting in an increase in the effective chain length. On escalators this wear is taken up by progressive movement outwards of the support carriage for the chain sprockets at the non-driving end, which in the case of London Underground is invariably the lower carriage. When this travel reaches its allowable limits the chains can be shortened by the removal of one or more steps with their associated lengths of chain. Ultimately however, the allowable wear is limited by the gaps between the steps which, for safety reasons, must not exceed 6mm to prevent the entry of shoe heels, umbrellas, or other passenger held objects. Within these criteria, the lives of London Underground escalator step chains varied between about 5 and 25 years, as shown in Figure 2. This is quite a major variation and the results were therefore analysed further to see if these variations could be explained. Some of the variations shown in Figure 2 arise from variations in the design of the escalators, and it is noticeable that the M series machines, which are of very robust design, tend to have the longest chain lives. The largest variation was shown by the LHDM type and the failure history of this type was therefore examined more closely, to see if the failure was due to wear out or some other cause.

A useful technique for determining the generic failure mechanism that is operating is shown in Figure 3. This uses the concept of the well known bath tub curve for failure rate variation with time in service to determine the failure mechanism. A more precise analytical technique for doing this is by Weibull analysis and when this was applied to the LHDM failure data as shown in Figure 4 it was clear that there were two different failure mechanisms operating. Step chain failures after more than 10 years in service were all as a result of wear, but those after less than 10 years were by some form of more random event. These escalators were identified and further enquiries revealed that this particular batch of escalators had had problems with the chain grease nipples becoming damaged by hitting parts of the structure, so that regreasing had been less effective. It was however a useful indication that effective chain lubrication was a very significant factor in chain life.

The need for step chain renewal as a result of progressive overall wear as shown in Figure 2 suggest a need for further improvement by achieving less scatter in the wear life and also ideally to increase the actual wear life to about 40 years to match a reasonable target life for escalator renewal.

This progressive overall wear of the chains was not however the major problem. The greatest operational difficulties were caused by the tendency of individual pins and bushes in the chains to become stiff or severely worn. Stiff pins and bushes can cause chain kinking and excessive sprocket carriage movement causing the escalator to trip and stop. An individually worn pin also increases the load on its adjacent pins and bushes, when entering the driving sprocket, due to the resulting pitch error. This then causes the adjacent pins and bushes to wear as well giving sections of chain with high wear, and these sections then need to be replaced. This situation has been observed in escalator operation.

The real problem organisationally however, is that if an escalator has several hundred pins and bushes in its step chains and each one can deteriorate in a random manner at any time, there is then no consistent and controllable chain life pattern to enable maintenance to be organised in a consistent and time effective manner. There was therefore a clear need to understand the nature of the failure mechanism in greater detail.

London Underground therefore supplied Neale Consulting Engineers with about 20 sample lengths of chain immediately after removal from service due to wear problems. These were then carefully dismantled and examined to investigate the nature and mechanism of the wear. The wear mechanism was found to be similar in every case and the wear arose from the direct rubbing together of the pins and bushes in metal-to-metal contact, resulting in adhesive wear. This gave rise to roughening and grooving of the surfaces, as shown in Figure 5, which indicates the progression of the wear and damage. Also when looking at a group of adjacent pins from one chain there is a general tendency for the wear to be more severe on the shorter hollow pins where the chain connects to the steps, as shown in Figure 6. This arises because the duty of these hollow pins is more severe, in that they carry more movement than the solid pins, due to the action of the chain, supported by the step wheels, when passing around changes of slope in the escalator track.

A factor in the operation of metal-on-metal surfaces in a small angle oscillatory contact is that any wear debris has difficulty in escaping. It can tend, therefore, to block the working clearance between the pins and bushes and thus prevent the entry of any fresh lubricant. The step chain pins and bushes on London Underground escalators each have an individual lubricant feed point which is lubricated at intervals as a regular maintenance procedure. The most commonly used lubricant is a semi-fluid grease containing graphite or molybdenum disulphide. However, the lubricant supply into the loaded area between the pins and bushes depends ultimately on the relative rotational oscillating movement between the two parts. If, however, the clearance is blocked by solidified wear debris, new lubricant cannot be supplied and the wear becomes progressively greater. This is therefore a mechanism capable of causing an unstable high wear rate on any individual pin and bush in an entirely random manner.

An interesting positive proof of this mechanism was obtained from a length of worn step chain removed from an escalator at Bank station. One of the pins on this chain had, for some reason, become loose in its sideplates and had, as a result, been rotationally creeping round in service, as shown in Figure 7. While this situation is undesirable in terms of the pin and sideplate contact, it does considerably improve the lubricant feed and distribution into the loaded contact area between the pin and its bush. The condition of this surface on the particular pin was very good, while on all the other pins in the same section of the chain there was severe wear and surface grooving.

Once this mechanism was understood it could be argued that an improvement might be obtained by more frequent and increased lubrication. This would however, not eliminate the problem, but possibly reduce it, at the expense of considerably increased lubricant contamination of the underside of the escalators. This was not acceptable because of the increased fire risk.

The next logical step, therefore, was to carry out a more detailed engineering analysis of step chain operation to determine any other viable engineering solutions which could give an acceptable long term wear life for the chains and in particular could eliminate the short term unstable wear behaviour of the many individual pins and bushes in the chains.

AN ANALYSIS OF STEP CHAIN OPERATION

The previous investigations had exposed a need for an improved bearing configuration for the pins and bushes on step chains. The available space does not allow enough room for conversion to rolling contact bearings, and in any case these do not usually give reliable performance on small oscillating movements because of the difficulty of maintaining a lubricant supply to the loaded contacts. They would also be appreciably more expensive. Some form of flexure connection could be possible but this would involve a totally new concept in chain design and a need for very extensive testing to ensure passenger safety. The best arrangement was therefore to continue to use a plain bearing design and to look for improved materials.

Plain bearings in rubbing contact depend for their satisfactory performance on operation within an allowable maximum contact pressure, and the amount by which they wear depends on the product of their bearing pressure and total sliding distance.

The family of London Underground escalators was therefore analysed to determine these various values and to calculate the expected wear rates of various plain bearing material combinations. For the majority of the escalators the peak bearing pressure on the chain pins was less than 25 N/mm^2 and the product of bearing pressure x the rubbing distance per year was in the range 2.5×10^{10} to 4.7×10^{10} N/m. These operating values are well within the performance range of modern reinforced plastic rubbing bearing materials, and the estimated wear life to an allowable maximum wear of 1mm per bush was of the order of 50 years, for some commercially available materials. It is of interest to note that these wear figures do not vary much with escalator rise height, because although the pin bearing loads reach higher values on high rise escalators, the number of rotational movements are less because the step chains spend a longer time operating in a straight line without bending.

An additional merit of the plastic rubbing bearing materials, is that after a small amount of initial bedding in wear, the wear rate is steady and consistent. They would therefore completely eliminate the unstable random wear and failure problems associated with lubricated steel on steel components as currently used. The plastic material would be used as a thin walled bush inserted within the steel bush and the pin would need to be made with a smooth non corrodable surface. This can be achieved by the use of pins either chromium plated or made from stainless steel.

These results indicated that it would be worthwhile to carry out a test programme to check the viability of this design in practice.

THE TESTING OF A VIABLE SOLUTION

In order to check the performance of the various possible material combinations it was considered desirable to test them on a rig before incorporating them in an actual escalator step chain. No suitable test rigs were available which could produce the required load and movement and take bushes of a representative size. A new test rig was therefore designed and manufactured and is shown in Figure 8.

A test programme was run on a number of different materials including various thin layer PTFE materials and also fibre reinforced resins impregnated with PTFE or other suitable materials. The condition of the test pins and bushes were inspected at intervals during the tests and the wear rates were measured. The optimum material in terms of performance and cost was found to be Railko NF22S which was a fibre reinforced and impregnated polyester resin, and could be used with a wall thickness of about 3mm. The wear rate in the rig indicated that the same material used in step chains would give a wear life of the order of 50 years and was therefore a viable solution. The test programme also covered the effect of contamination, which is likely to occur in service due to construction work, and the materials were shown to be tolerant to hard debris particle contamination.

Two complete sets of escalator chains were therefore manufactured to this specification and installed in two London Underground escalators. The lower carriage movements on these escalators was regularly monitored to check the wear performance, and portions of chain were removed for detailed examination after 1 year and 4 years in service. The step chains have performed very well and have given wear rates that are closely in line with expectations, for chain lives of the order of 50 years. This life expectancy is indicated by extrapolation of wear rates experienced during a 4 year in-service trial. It was necessary for the trial period to extend over a number of years to ensure that life predictions were accurate, but more importantly to ensure that there were no other factors affecting chain performance or chain safety, such as creep of the polymer bushes, contamination by abrasive particles or contamination by water.

Monitoring of the trial escalators is continuing in order to further increase the accuracy of the life prediction, however LUL now have sufficient confidence in the improved performance of the polymer bush chain to be introducing the new chain design to more escalators within the LUL infrastructure.

CONCLUSIONS

This paper has described a very successful procedure for improving the reliability of an essential part of a railway infrastructure. This has involved a close working relationship between the full time staff of the railway system and a small group of specialist engineers with a detailed understanding of the technology involved. The service experience was studied to identify a key problem, which was then approached methodically to determine the engineering issues involved and derive a practical solution giving greatly increased reliability.

The development has the potential for giving major service improvements and reduced costs on London Underground escalators. It has also opened up major new commercial opportunities for UK material and component suppliers and the developments have already been adopted in other countries such as the Paris Metro and the New York Subway. It is hoped that it can provide a model for further development work on other parts of railway systems, to approach a fault free infrastructure.

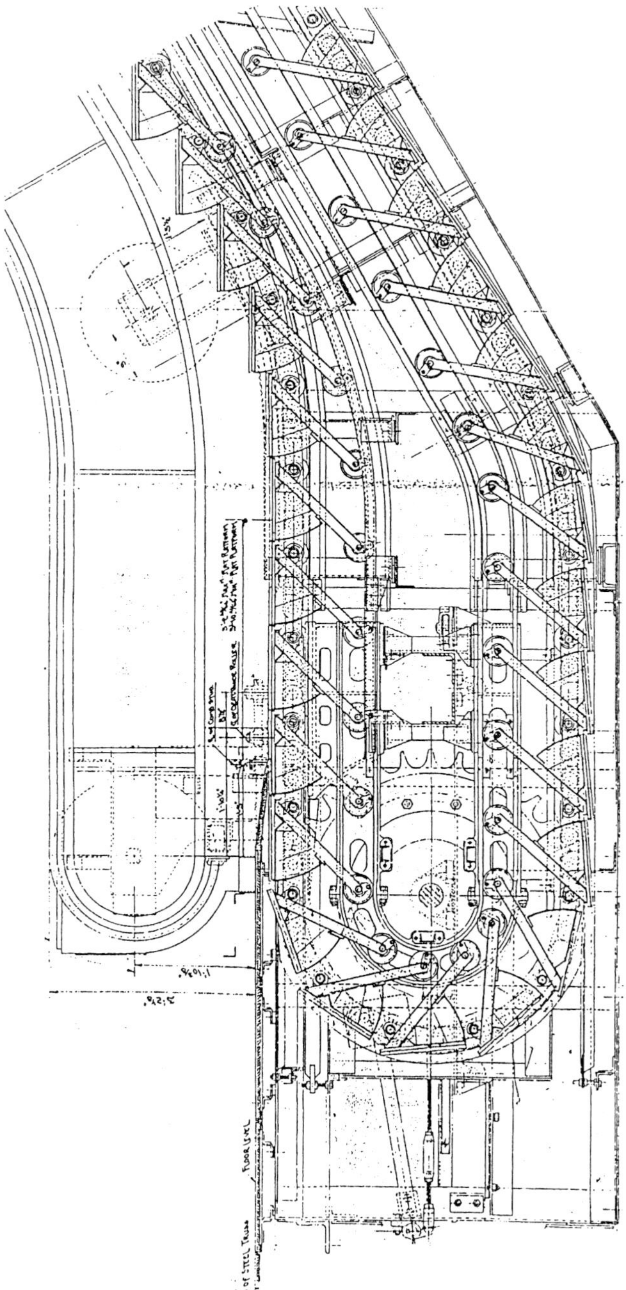

Figure 1 Cross section of an escalator

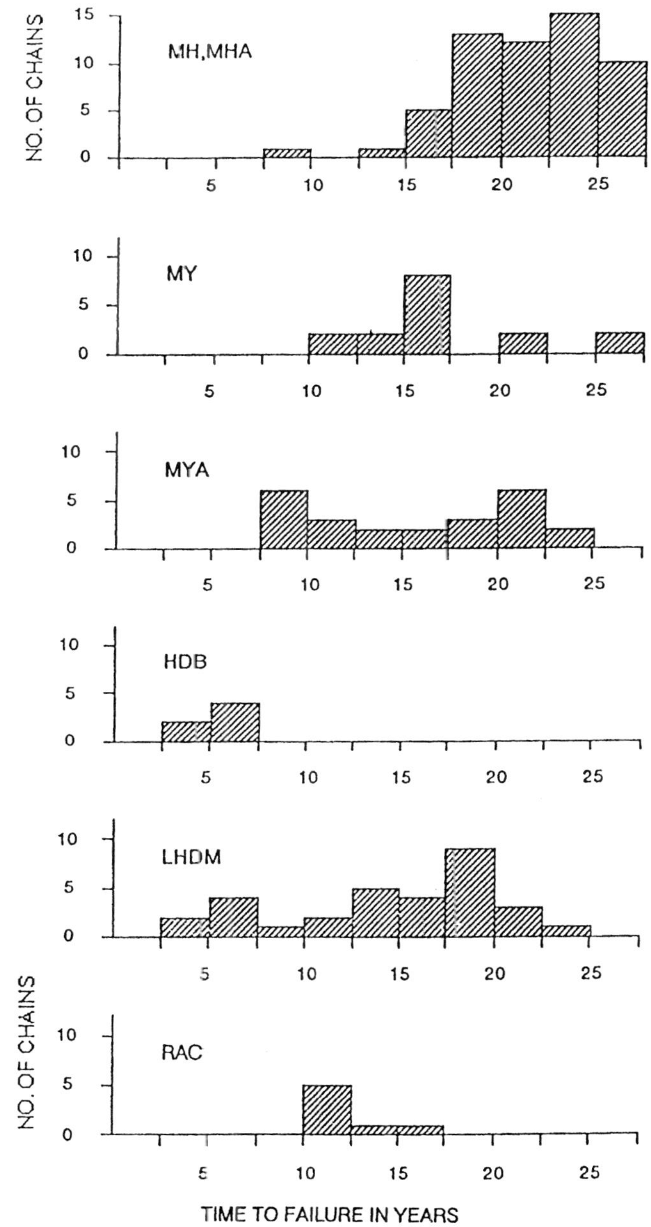

Figure 2 The failure patterns of the step chains of the various types of escalators

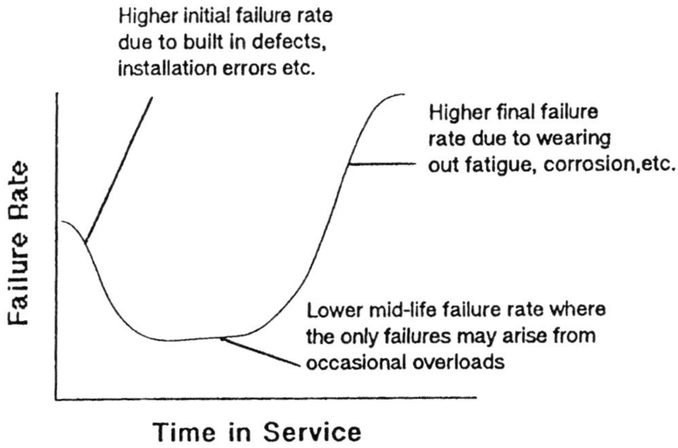

A typical bath tub curve for the failure rate of engineering components.

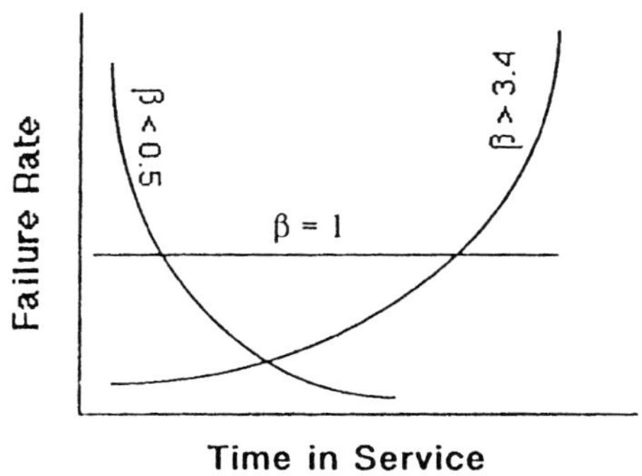

The portions of the bath tub curve reproduced in Weibull analysis by various values of the shape parameter β

Weibull Shape Parameter β	Engineering Significance	Human Equivalent
0.5 or less	Poor manufacture or assembly	Infant mortality
1.0 approx	Random overloading	Motor cycle accidents
3.4 or more	Wear out	Old age

Figure 3 A typical bath tub failure rate curve and the associated meaning of the shape parameter β in Weibull analysis

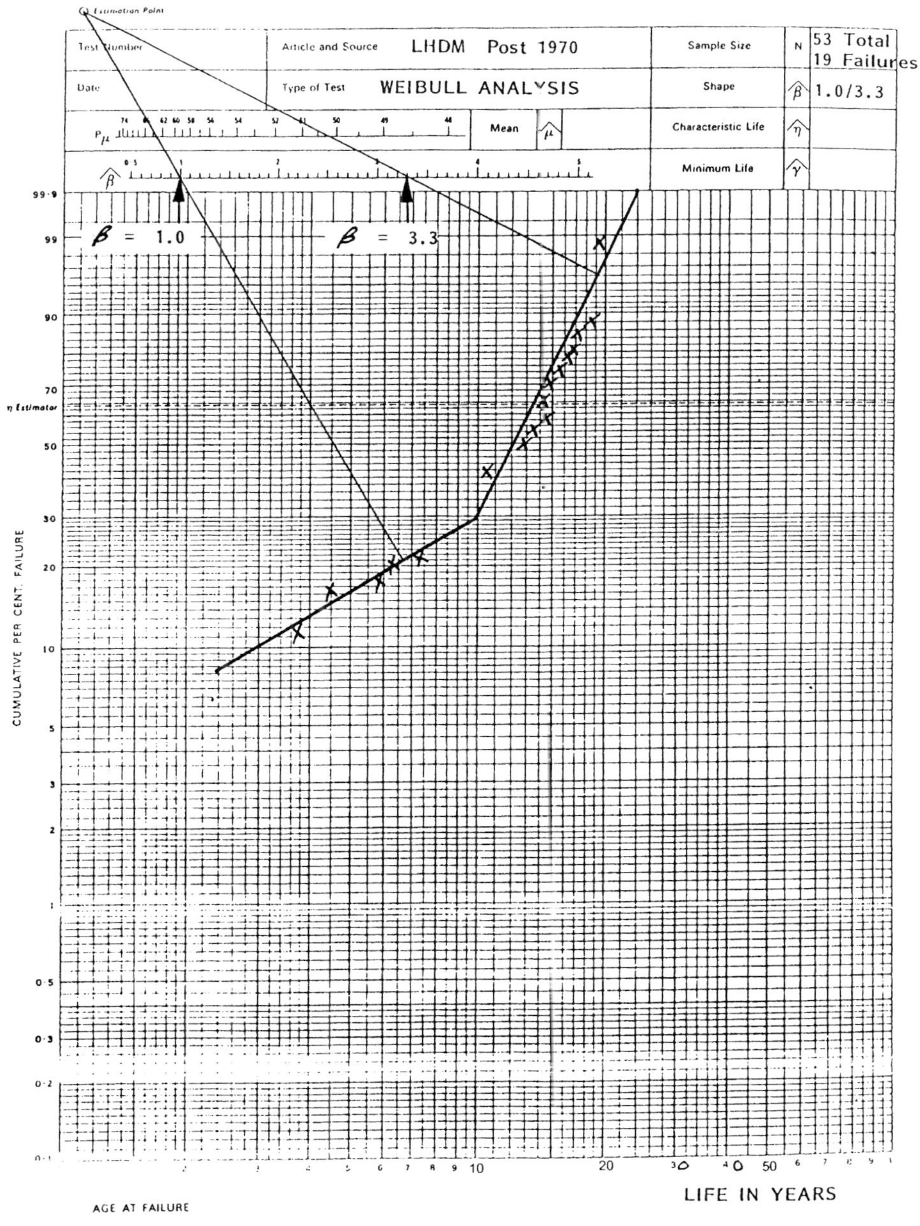

Figure 4 The life pattern of LHDM escalators showing two generic failure types corresponding to β = 1.0 and β = 3.3

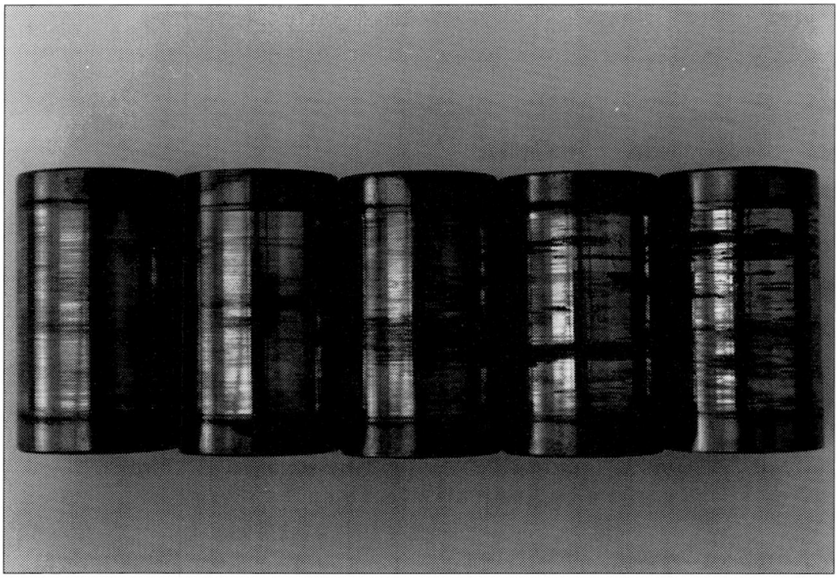

Figure 5 Pins demonstrating the rang of severity of surface damage

Figure 6 A sequence of pins with more wear on the hollow (shorter) pins

Figure 7 A MH pin which has been rotating in its sideplates (left) compared with a typical pin from the same chain

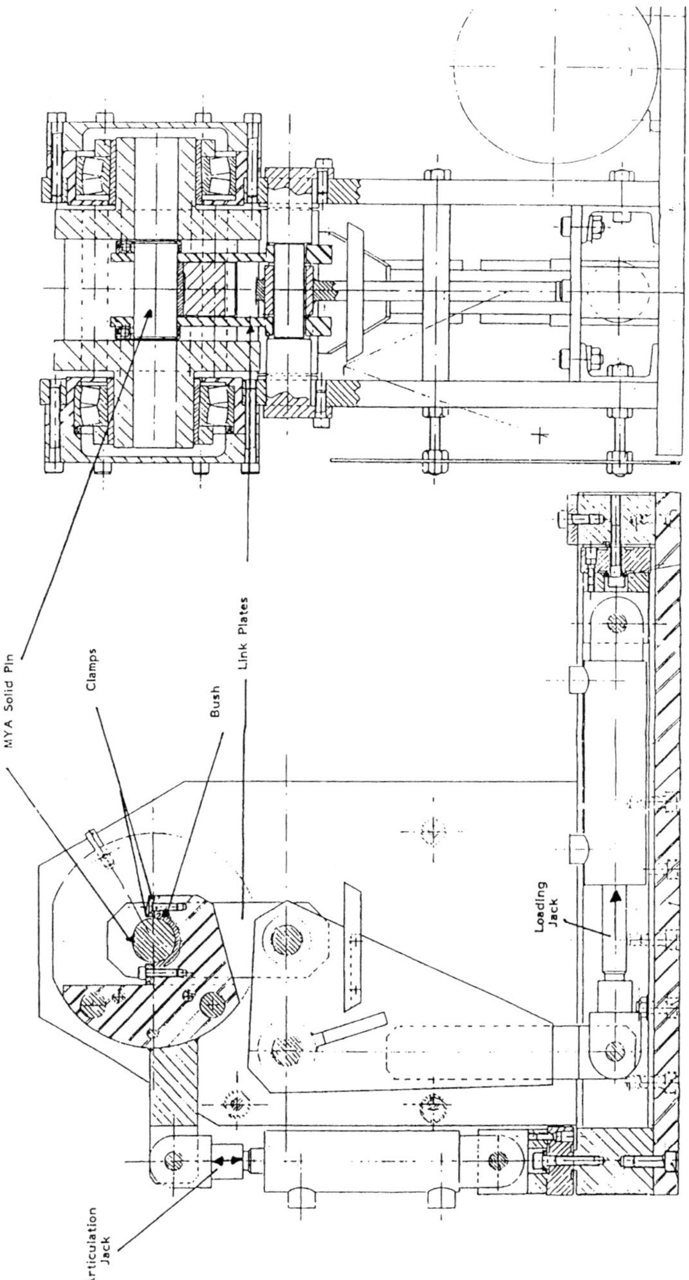

Figure 8 Arrangement of the NCEL designed pin and bush test rig

C570/020/99

Fault-free infrastructure – escalator maintenance

Y-S AU
Mass Transit Railway Corporation, Hong Kong

ABSTRACT

Since the passenger operation of the Mass Transit Railway (MTR) in 1979, the total number of escalators in the Urban Railway Lines has increased to 462 units in 1998. These escalators serve a daily patronage of 2.3 million and form an essential service for normal passenger flow and emergency evacuation. Maintenance of these escalators has been contracted out since day one of operation. Similar to other Metros, the Corporation relentlessly seeks continuous improvement in Safety, Reliability, Recovery, and Efficiency to meet the high demand of passengers. However, some of these objectives may not be aligned with the Contractor who can be more contractual and cost sensitive.

This paper introduces an integrated approach for escalator maintenance and shows how a partnership can be established with the service contractor to achieve these objectives. The main purpose is to work towards a common goal, sharing of information to achieve a high standard of performance, a higher profit margin to the contractor and less cost to the railway.

The paper includes the main features of an Integrated System and the achievements obtained since implementation from 1993 onwards. The author also identifies the main challenges and the key success factors of the system and lessons learned during its implementation.

1. INTRODUCTION

The Mass Transit Railway Corporation has served Hong Kong since 1979. In 1998, the average daily patronage reached 2.3 million. There are now 462 escalators serving the Urban Line. Escalators form an integral part of the transport system in delivering passengers to and from city street levels to station concourses and platforms. The escalators operate for 19 hours a day and 7 days a week. They form an essential part of the daily operation, for normal passenger flow and emergency evacuation of the MTR stations. Breakdown of escalators may cause severe disruption and passenger accidents. As a result, failure elimination measures need to be progressed at all times.

Maintenance of escalators requires effective utilisation of manpower and an effective management system over their asset life. The additional requirements for Safety and enhanced Reliability have introduced further constraints. According to MTR experience, the integration of an Effective and Efficient performance and condition Feedback mechanism into the Asset Life cycle is the most feasible strategy for eliminating possible escalator failures. Through this, escalator Safety, Reliability, Recovery and Efficiency can be achieved.

The implementation of an Integrated Escalator Management System on MTR has followed this strategy and has demonstrated a high degree of success on failure mitigation. Since the establishment of the Lift & Escalator (L&E) Team and the implementation of the system in 1993, escalator proportional failure rate has reduced and availability has improved continuously. The system is driving towards a Fault-free Maintenance scenario. Full co-operation from both MTR and the maintenance Contractor, clear objectives and performance target setting, systematic and logical planning on preventive maintenance, and partnership between MTR and the maintenance Contractor are the key driving forces for success. In brief, the Integrated Escalator Management System includes all of the life cycle stages of Design, Installation, Operation and Maintenance, and Asset Renewal (Fig 1). Faults are identified along the path. Feedback is consolidated from all stages for Continuous Improvement. The following subsections highlight the essential elements.

2. INTEGRATED ESCALATOR MANAGEMENT SYSTEM

2.1 Design

2.1.1 Design Specifications

Initial escalator design requirements are stipulated in the Particular Technical Specification. Following changes in the Statutory requirements, Design Standard, Station Environment, Customer requirement, and as a result of Maintenance Feedback and Asset Renewal Feedback, the original Particular Technical Specification was revised with enhanced features and performance criteria with the objective of designing defects out rather than designing maintenance in.

The New Particular Technical Specification serves as a common base for future MTR development or existing escalator modification programmes. This is a standardized and consistent specification and the particular benefit of it is the inclusion of Feedback from Maintenance and Asset Renewal stages. Previous design deficiencies in respect of Safety, Reliability, Recovery and Efficiency can be examined thoroughly and reviewed with feasible solutions for continuous improvement.

A Hazard Registration process is applied for evaluating the risk level and identifying the possible improvement action. The Particular Technical Specification will be enhanced accordingly. For the older installations, modification will be carried out to enhance their performance physically. The modification is referred to as the Capital and Revenue Works of the MTR. For new installations, improvement eliminates the recurrence of past faulty experience.

2.1.2 Contract Strategy Determination

The maintenance stage could typically be eight times the duration of the project stage for the escalator life cycle. The maintenance cost is likely to increase still further when the maintenance duration is planned in excess of the Optimum Asset Replacement Life. Maintenance cost tends to escalate as time goes by. Maintenance also requires extensive planning and coordination efforts to reduce maintenance costs against time. For reducing planning and coordination effort, MTR has adopted a contract out maintenance strategy instead of in-house maintenance. Currently, MTR concentrates on supervision and contract management while the maintenance Contractor concentrates on the fieldwork and service recovery.

With a view to influence the design and to reduce maintenance costs, the MTR has integrated the maintenance contract requirement into new project tenders. This forms part of the Tender Evaluation, which is welcome by both MTR and the Contractor. This also supports the growth of a supplier maintainer practice in the Hong Kong escalator industry. When a Contractor is able to supply and maintain their product under a continuous contract, it should be able to attain a better financial position. The particular benefit exists in asset management; the integrated contract approach and the subsequent long-term relationship with the Contractor allows the Contractor to perform maintenance work with comprehensive quality, longer loyalty and higher self-initiative. The Contractor's ability to provide staff training and maintenance support at an early stage could produce more competent maintenance staff and logistic backup. Then preventive maintenance work performance can be assured at a later stage and the possibility of escalator fault minimised. Further, transparent costing data allows better financial arrangements and budgeting for both parties.

In view of the benefits, the inclusion of maintenance requirements into the project tender for new installations at the initial tendering stage is strongly recommended. The benefit has long term effects and the resultant escalator quality is more cost-effective over the longer term.

2.2 Installation

To ensure the Contractor complies with the Design Specification, high quality inspection on installation work is necessary. Otherwise, hidden problems from the installation stage will appear later at the maintenance stage and lead to unnecessary failure and poor escalator reliability. These will then require exhaustive effort and high cost to rectify.

In order to rectify any deficiencies early for the sake of long term interest in escalator performance, a team of competent staff is necessary to audit the installation work during the Testing and Commissioning (T&C) phase. The MTR's L&E Team has been chosen for this task as they are more suited than their project colleagues at this stage in the project. The L&E Team members were highly experienced in the escalator maintenance industry, before joining the MTR. Their duty is to supervise and monitor escalator performance throughout the Operation and Maintenance stages. The L&E Team could apply their latest maintenance experience against installation deficiency. Furthermore, first-hand T&C information can be applied in monitoring escalator performance particularly during the Defect Liability Period.

2.3 Operation and Maintenance

2.3.1 Maintenance Strategy Evaluation

Clear objectives set for escalator maintenance provide MTR and the Contractor with Common Goals and Consistent References throughout the Operation and Maintenance stages. As Safety and Customer Orientated Services are the primary concern of MTR, the quality of escalator maintenance and operational reliability have been continuously improved to minimise the possibility of escalator fault. The disruption to passenger service can be kept to a minimum because MTR can work with the Contractor in Partnership, Maintenance Records and Information is shared so as to facilitate speedy communication and response in problem solving. All these essential elements have been addressed in the Maintenance Strategy formulation. Finally, compromise between the captioned objectives and costing is required in order that the Maintenance Strategy can be converted into Maintenance Contract in a practical and cost effective manner.

2.3.2 Maintenance Contract

In line with the Maintenance Strategy, the Maintenance Contract will be formed under three main headings Scope of Work, Performance Targets and Contract Administration.

2.3.2.1 Scope of Work

The Contractor under the terms and conditions of a 5-year comprehensive renewable maintenance contract carries out routine escalator maintenance. All labour and material costs are included in the maintenance contract. The contract meets the statutory maintenance requirements as stipulated by the Hong Kong Government.

Preventive Maintenance frequency and scope have been enhanced by MTR. The maintenance schedule of the escalators, covers the entire statutory requirements, is based upon time intervals and consists of daily running inspection, weekly routine maintenance, and monthly service, 3-monthly, 6-monthly and yearly overhaul.

The daily running inspection aims at identifying potential hazards to passengers and to highlight potential problems that would cause disruption to escalator service. Weekly routine maintenance is the preventive maintenance and includes checking of all the safety devices, cleaning and lubrication. Monthly service is aimed at minor repairs and adjustment to the escalator parts. 6-monthly service provides an opportunity for planned replacement of consumable items, adjustment of major items, functional test of all safety devices, thorough cleaning and measurement of the tolerance of major items.

2.3.2.2 Performance Targets

The initial contract documents were derived primarily from contemporary commercial contracts with loose performance targets specified. In the light of past maintenance experience, more stringent control and performance clauses have been included, as appropriate, during the contract renewal negotiations.

2.3.2.3 Contract Administration

The contract adopts a 3-level Maintenance Contract Administration (Fig 2) in order that partnership can be exercised between MTR and the Contractor at three organisational levels. The 3-level Maintenance Contract Administration allows maintenance problems to be solved at the most suitable administrative level, and facilitates effective communication and appropriate authorisation. Hence, escalator faults caused by poor decision at inappropriate administration level can be avoided.

Level 1 - the Chief Engineer or his delegate which involves the development of a Strategy and Contract Negotiation. He also acts as the arbitrator when there is a conflict or disagreement at operational level. Level 2 - the Electrical and Mechanical (E&M) Engineering Section Head, for regular Contract Administration with the contractors. Review meetings on agreed performance and service targets are held, usually on a monthly or quarterly basis. Level 3 - the E&M Section Engineers and the L&E Team, for the day-to-day Operation Administration. They keep in close contact with the contractor's maintenance staff, follow up on all the work completed, quality control, verify corrective maintenance, root cause finding and subsequent remedial action or Third Party Damage justification.

2.3.3 Maintenance Tasks Scheduling

2.3.3.1 Escalator Classification

The loading for the escalators varies according to different station environments and by time of day. Classification of escalators into different groups according to operational requirements is essential for the effective preventive maintenance planning and station crowd control. The prime aim is to minimise interruption to passenger service. Careful scheduling and high quality Preventive Maintenance on the super critical escalators helps to reduce the possibility of failure, even though the escalators are being used the most heavily. Contractors can then allocate their resources more effectively with reference to escalator criticality. In 1990, the MTR escalator classification system was established with Super-critical, Critical and Non-critical categories. Each category as different customer service targets and service requirements.

2.3.3.1.1 Super-critical Escalator

Super-critical escalators are those at the busiest locations without an alternative route for passengers such that there is no redundancy, no adjacent back up staircases, carry most passengers and tend to have a very long rise. Such escalators cannot be released at any time during traffic hours, even for the yearly overhaul under statutory requirement. This has to be scheduled in the shortest possible time with 24 hour shift working. In the event of escalator failure and suspension of the service, the Station Controller must take immediate action to divert the passenger flow, perhaps close an adjacent entrance and arrange to resume service as soon as possible. The expected failure rate must be minimal in comparing with other classifications.

2.3.3.1.2 Critical Escalator

Critical escalators are those which are as busy and heavily used as super-critical, but they are adjacent to other units or a fixed staircase, which can provide an alternative route for passenger flow. They are not as critical as the super-critical units; but if they fail they would dramatically affect passenger convenience, especially during peak hours. Regular maintenance work is only allowed during non-traffic hours. Annual overhaul, which requires 24 hour shift working, can be scheduled with proper advance announcement to the public. Although they are not super-critical, once failed they are likely to cause the remaining escalators to become Super-critical.

2.3.3.1.3 Non-critical Escalator

Non-critical escalators are those that will create a lesser impact to station operator's if they are out of service. There will be sufficient alternative routings to divert the passenger flow. Both regular maintenance and annual major overhaul can be carried out during traffic hours, normally off peak.

2.3.3.2 Customer Service Targets

To provide a common goal and escalator performance objectives for the Contractors to plan and carry out maintenance work, Escalator Customer Service Targets have been established. The MTR and the Contractor accept the targets which will then be included in the Contractor's maintenance contract for common contractual reference. Targets are set for Escalator Availability, Proportional Failure Rate, Total Response Time and Total Recovery Time.

2.3.3.2.1 Escalator Availability

It is defined as the % of the station operating hours the escalator is available for services. The total Operating Time is calculated from the basis of 19 hours of escalator operation per day. While downtime from Preventive Maintenance and Corrective Maintenance are combined as the Total Service Down Time (Fig 3).

Availability = (Total Operating Time) / (Total Operation Time + Total Service Down Time) * 100%.

2.3.3.2.2 Proportional Failure Rate

It is the ratio of the total number of failures to the number of escalators for a month. As the number of escalator units changes over time, an examination of failure performance by counting the number of failure needs to be normalised. Therefore the proportional failure method is adopted for consistent and long term comparison of escalator performance, by station, line and manufacturer. This benchmarking also highlights critical components and fault trends for early remedial action, which is part of MTR's fault elimination approach (Fig 3).

2.3.3.2.3 Total Response Time

It is the time that elapses between the reporting of a fault and the maintenance personnel arriving at the location of the faulty escalator (Fig 3).

2.3.3.2.4 Total Recovery Time

It is the time that elapses between the reporting of a fault and the affected escalator regaining its ability to perform all of its required functions (Fig 3).

The objective being a quick response and minimum repair time.

2.3.3.3 Monthly Escalator Maintenance and Capital & Revenue Work Schedule

During traffic hours, maintenance work will cause downtime and service interruption. Because of the classification and customer service target constraints, maintenance planning and work scheduling becomes critical. Hence, Monthly Escalator Maintenance and Capital & Revenue Work Schedules have been established in MTR. The schedule allows the Station Controller, the Contractors and the L&E Team to have a common overview on all Preventive

Maintenance and Capital & Revenue works, so as to ensure a satisfactory completion rate of maintenance activities. Sufficient working time can avoid incomplete work quality that may course premature failures.

2.3.4 Maintenance and Operation

2.3.4.1 Work Quality Assurance

The Maintenance and Operation phase covers almost 85% of an escalator life. Maintaining the escalator performance up to the specified targets requires sufficient quality control and assurance. Work must be done correctly first time thus avoiding rework later. Good work quality is the key factor for the success of minimum failure rate. The scope of activities involved in Preventive Maintenance are addressed in the Maintenance Contract sub-section.

In addition, in-depth investigations, and root cause analysis for all Corrective Maintenance work can avoid recurrence and repetitive failure. In the long run, reliability is improved and failure rate reduced. A serious investigation attitude can also foster a prompt recovery thus minimise service disruption. Failure records are an important database for monitoring the asset condition. For example, mechanical faults in step assembly contributed 8% of the total fault record in 1998. With this information, particular preventive measures on step axles and step running clearance can be progressed against those identified components. The Technical Life of escalator components can be compared with their Design Life. Furthermore, proper Capital and Revenue Work schedules can be developed to enhance the component condition if preventive maintenance is unable to rectify the condition.

2.3.4.2 Electrical and Mechanical Management Information System

The E&M Management Information System captures all the essential data of Preventive Maintenance and Corrective Maintenance works on each escalator. Corrective Maintenance data is important for fault elimination and condition monitoring. The data captured includes symptoms, causes, components affected, remedial actions, and the time element of Contractor's work. Summaries of Escalator Availability, Proportional Failure Rate, Response Time and Recovery Time will be consolidated for meeting discussions and specific analysis. Currently, Contractors are responsible for data input of root causes and actions taken in the Corrective Maintenance Work. This process allows a minimum amount of reporting and provides maximum information for reference. The L&E team will subsequently verify the data. The data is shared and Contractors can retrieve their own performance record for review. The joint efforts by the Contractors and the L&E on Management Information System data input and maintenance show another benefit derived from a partnership approach.

2.3.4.3 Contractor

The Contractor performs a key role during the Maintenance and Operation stage. The efficiency and effectiveness of the contractor's organization and procedures will determine the degree to which they can meet MTR's requirements. Interactive communication and a good partnership atmosphere is essential to foster smooth implementation. In addition to customer service target setting and common goals, the Contractor should review their workforce capability and financial commitments before agreeing to a maintenance contract with MTR. This should eliminate the possibility of entering into a maintenance contract with insufficient cost recovery, which would eventually lead to unsatisfactory work and high failure rate.

The Contractor's maintenance staff has to be technically competent for the maintenance works. They should be well trained with sufficient skill and job knowledge to discharge their duties. Particular concerns are on thorough Preventive Maintenance, Corrective Maintenance and technical investigations. If they cannot perform their duties competently, extra effort will be required from either the Contractor or the MTRC or both. This is not cost-effective and makes it more expensive for MTR to monitor the maintenance performance.

2.3.4.4 Passenger

Failures due to Passenger actions are considered to be a third party cause of escalator failure and have no connection with maintenance contract administration. The passengers need regular education to improve their behaviour and enhance their safety awareness when using escalators. This is best done when the passengers are at a young age. The earlier the passengers are educated, the deeper the safety awareness that can be built up. Hence, MTR's recent effort has been geared to the youth and children. Regular reinforcement of the safety message is achieved through different media with safety posters, safety campaigns, TV programmes, safety ambassadors etc.

2.4 Asset Renewal

The recent MTR Escalator Replacement Strategic Review concluded that the optimum escalator replacement period to be 25-30 years. As the railway has now been in operation for over 20 years, it is an appropriate time to consider and prepare for the replacement activities. It is anticipated that escalator failure rate will deteriorate with the passage of time and continual usage. The escalator Design Life is less relevant and an annual escalator condition survey has been established to determine each escalator's remaining Technical Life. A deterioration Index can then be applied to identify the priority for fleet escalator or major component replacement before the overall escalator failure rate deteriorates to an unacceptable level. This also provides feedback and reference for future Design and Maintenance stages.

As more components need replacing, maintenance costs will tend to increase with extended escalator life. To obtain the optimum year for escalator replacement whilst maintaining the escalator performance, a Life Cycle Costing approach has been applied to determine the optimum, most cost effective replacement time. With the assistance of performance trends on Proportional Failure Rate and Availability, and information on present condition a strategy to Do Nothing, Refurbish, Re-build or Replace can be developed. Of particular concern for the MTR, is the large number of escalators (462) and rebuilding 20/year would take more than 20 years by which time the condition could further deteriorate.

Financial evaluation of the options available, using Cost Benefit Analysis, Net Present Value or Internal Rate of Return techniques, will lead to a decision on the most cost-effective strategy for asset renewal.

3. LESSONS LEARNED

3.1 Main Challenge

To help the Contractors meet the MTR's requirements of minimum cost with maximum performance is not an easy task. Contractors in the maintenance industry are all looking to improve their profit margin with minimum resource input. It is important to foster a partnership with Contractors to eliminate waste by both parties, which has been the main challenge in the past and will remain so for the years ahead.

3.2 Successful Factors

The integration of condition and performance Feedback into the escalator maintenance strategy, to achieve continuous improvement is the main objective of the Integrated Escalator Management System. The key success factors are Common Goals, Safety, Customer Orientated Service, Reliability, Fault Free Approach. Partnership, Sharing of Information, Competent L&E Team, a Win-win attitude and Minimum Cost are all essential elements that need to be reviewed and addressed when implementing the System.

3.3 Achievements

3.3.1 Reduce Third Party Damage Cost

As a result of continuous improvement under the Contractor's maintenance and MTR's supervision, the root cause of each failure is identified. be it equipment/component failure or maintenance deficiency. In each case, the station staff will log the symptoms before notifying Contractors for checking or remedial work. For major incidents, or a passenger accident, the MTR's L&E Team will join the inspection, and costs are borne by the Contractor or MTR as appropriate. The Contractor seldom argues about the root cause and costs. Hence, claims by Third Parties on MTR are kept to a minimum and under control, which is also reflected as a lower-cost in our budget.

3.3.2 No Fatality or Serious Injury

With a good partnership between both MTR and their Contractors, and a high standard of maintenance through the 19 years of escalator operation MTR has carried billions of passengers, with an average 4 escalator rides/pass and there has been no fatality or serious injury as a result of negligence of MTR or equipment fault (Fig 4). Most accidents that do arise are as a result of passenger behaviour (such as improper use, running, sitting on escalator, loss of balance, unwell, drunk, etc) and MTR continues to enhance passenger education through posters, school talks, video shows, and escalator safety campaigns.

3.3.3 Availability Improvement

Escalator Availability is a good condition indicator and reflects the time spent on both Preventive Maintenance and Corrective Maintenance. Generally, the ratio of Preventive Maintenance (PM) downtime against Corrective Maintenance downtime is about 4 to 1. Strict control of PM is essential for maintaining a low downtime rate and high availability.

MTR and the Contractor work closely to formulate the Monthly Escalator Maintenance and Capital/Revenue Work Schedule with the consensus of all parties concerned, including Station Managers, MTR's L&E Team and the Contractor. The Schedule streamline the monthly PM workload as well as the downtime at stations. It minimises the possible monthly downtime fluctuation throughout the year and serves as a common reference and target for all parties concerned. Amendments and updates are issued to all parties concerned to accommodate changes. A Summary (Fig 5) of the Escalator Availability shows the improving trend of escalator performance between 1993 and 1998.

3.3.4 Failure Rate Improvement

The number of escalators is continuously increasing with new station developments and new railway extensions. The proportional escalator failure rate target, number of failures/escalator/month has been adopted as a common basis for comparing different maintenance Contractors/escalator performance on a monthly basis. With the assistance of the E&M Management Information System, repetitive failure and high failure rates can be easily identified for Contractors' attention and immediate remedial actions. The database created in the Electrical & Mechanical Management Information System over the years enables failure trend analysis, and also highlights the critical failure components for design review and enhancement. The proportional escalator failure rate has steadily been reduced by almost 30%, from 0.3 failures/escalator/month in 1993 to 0.21 in 1998 with the ultimate goal of zero defects (Fig 6).

3.3.5 Successful Escalator Supervision Transfer

Since 1993 the MTR L&E Team size has been reduced from 23 to 12 with the transfer of performance monitoring from the L&E Team to the Contractor. This reflects both the L&E Team's competence and the Contractor's work quality improvement with ISO9001 certification. This transfer of supervision reduces the MTR overheads and helps the Contractors handle their work more competently and thus reduces the escalator failures in the long run.

4. CONCLUSION

4.1 Escalators built with the latest technology are still containing major mechanical components, which are subject to normal wear and tear. An effective maintenance system is essential for good performance, and continuous improvement could be achieved though it might not be possible to be totally fault free.

4.2 The Integrated Escalator Maintenance System described supports good performance with continuous improvement, which require attention to the full life cycle of the asset and the partnership with the Contractor.

4.3 At the design stages, effective feedback on asset performance of existing units helps to reduce future escalator failures. Inclusion of a maintenance agreement in the main tender would enable better cost control of the escalator maintenance.

4.4 If maintenance staff are involved early in the testing and commissioning, deficiencies would be identified and will avoid costly rework afterwards.

4.5 During the maintenance stages, a long-term partnership with the Contractor and continuous improvement mechanisms built in the maintenance contract are key success factors. The Customer Service Targets serve as the common goals and objectives for performance monitoring. The adoption of a common maintenance management information system between the Railway and the Contractor assists to simplify communication between the two parties.

4.6 Whilst every means is applied to extend the life expectancy of the asset, an effective asset replacement strategy would be required. Life cycle costing and financial evaluation together prove to be an effective means to determine the optimum replacement time and cost.

5. REFERENCE

1. **Yan-Shu Au**. Heavy Duty Escalator - A Strategic Review Process for Escalator in Hong Kong Mass Transit Railway. *Dissertation for The University of Sheffield,* 1998.

Fig 1

INTEGRATED ESCALATOR MAINTENANCE SYSTEM

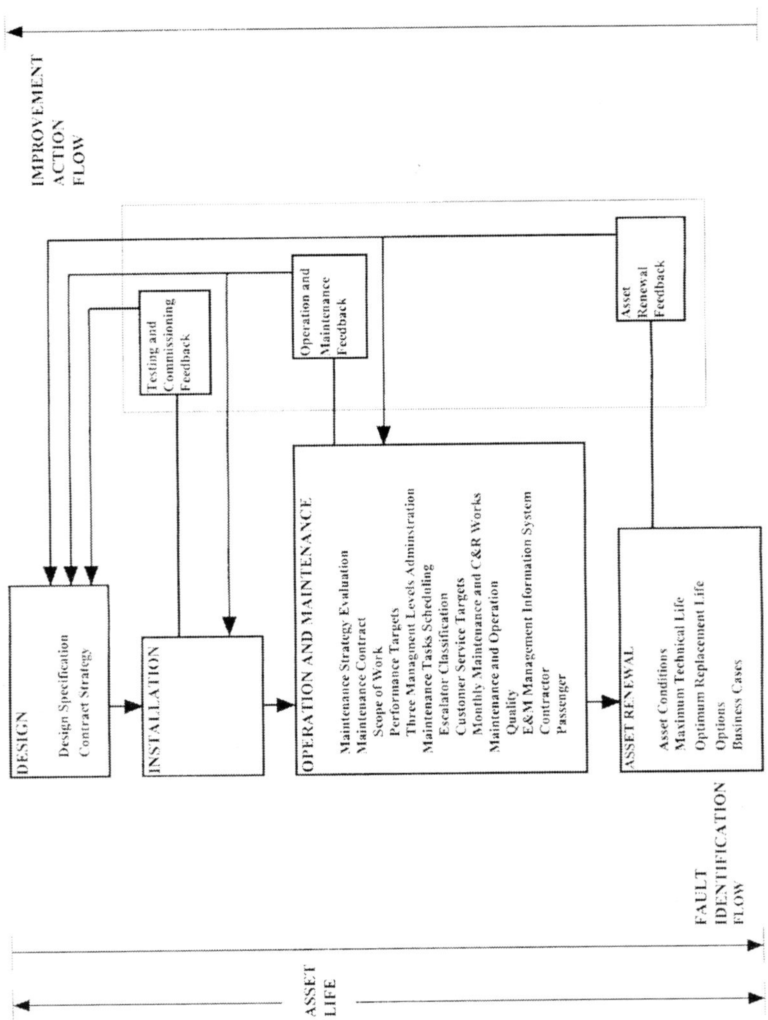

Fig 2

Three Level Maintenance Contract Administration

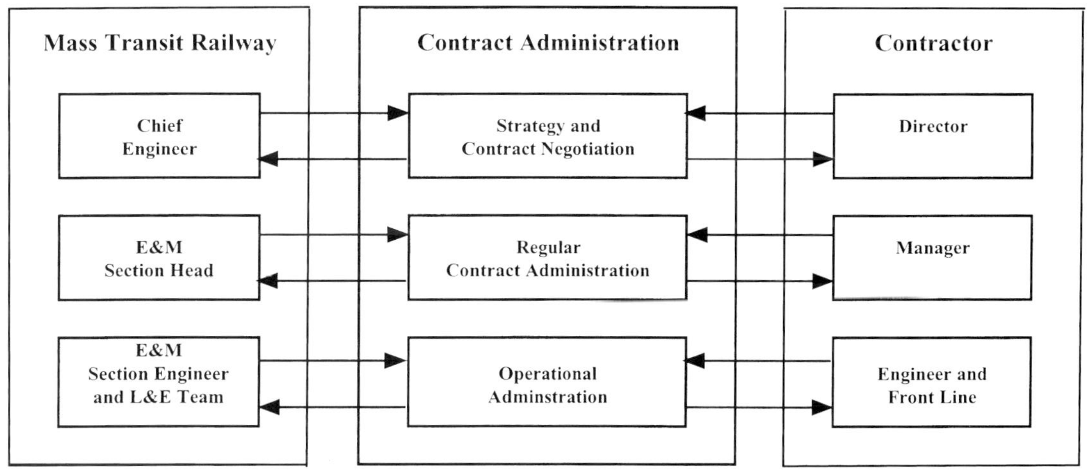

Fig 3

Monthly Customer Service Target Summary for 1999

Escalator Classification	Availability (%)	Proportional Failure Rate (Failure per unit)	Total Response Time (Hours per CM Job)	Service Down Time (Hours per CM Job)
Supercritical	99.800	0.25	0.50	1.00
Critical	99.200	0.30	0.75	1.50
Non-critical	99.100	0.30	1.00	1.75

Remark: CM = Corrective Maintenance after Failure

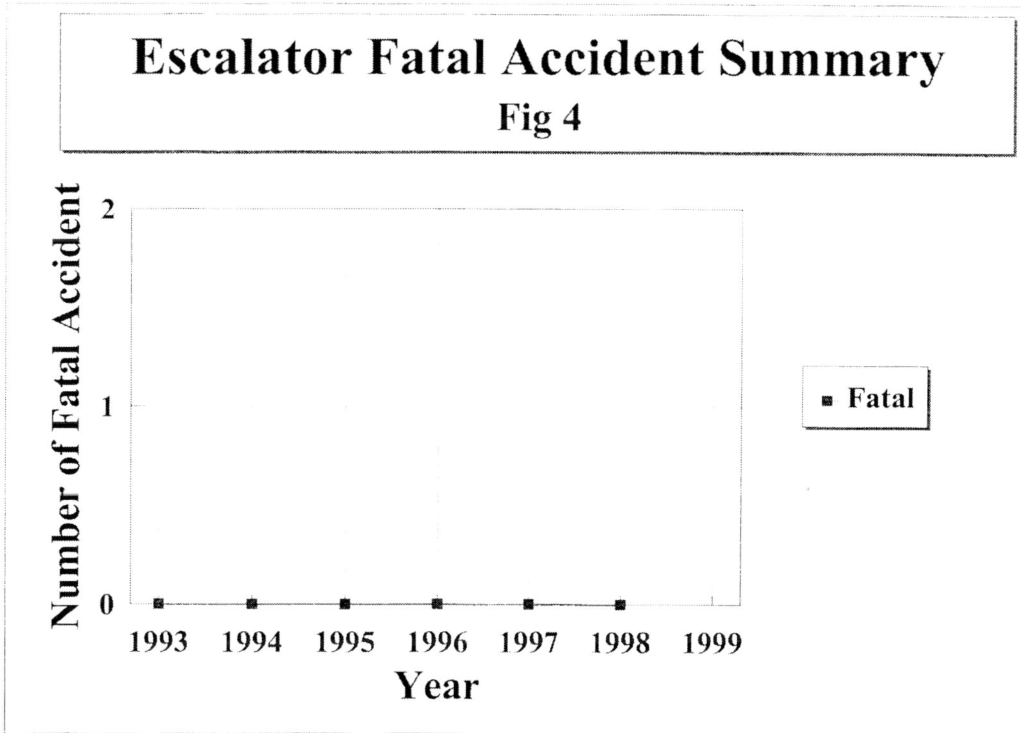

There have been no fatality in 19 years of operation, due to negligence of MTR or equipment fault.

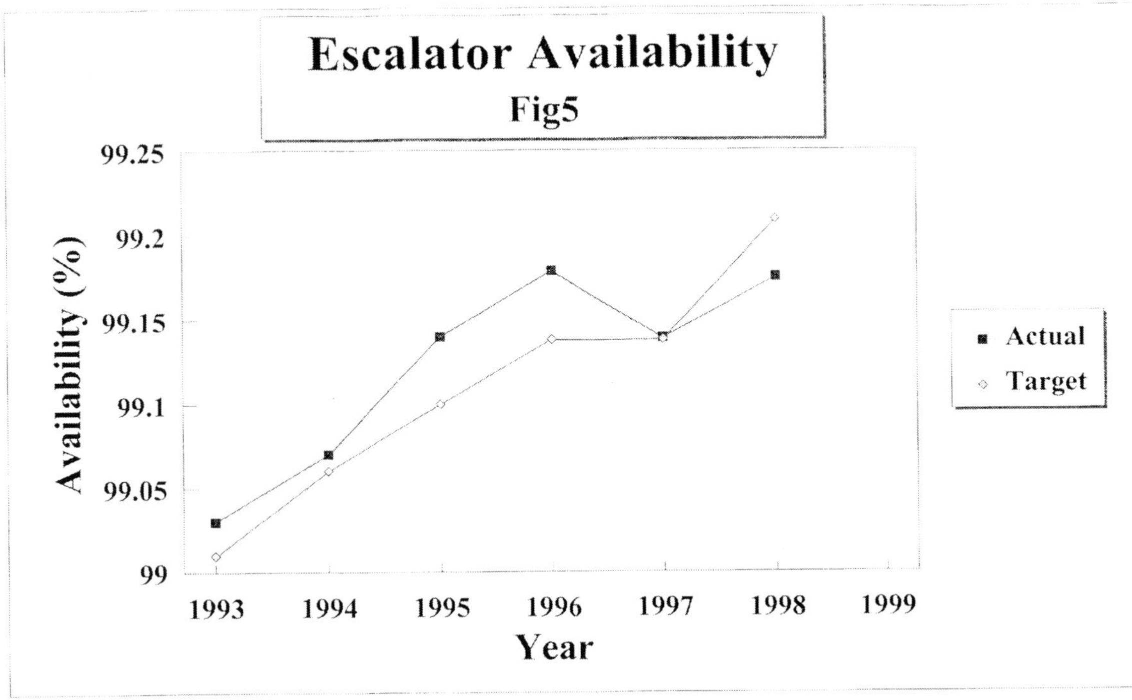

Availability
= (Total Operating Time) / (Total Operation Time + Total Service Down Time) * 100%

Escalator Proportional Failure Rate

Fig 6

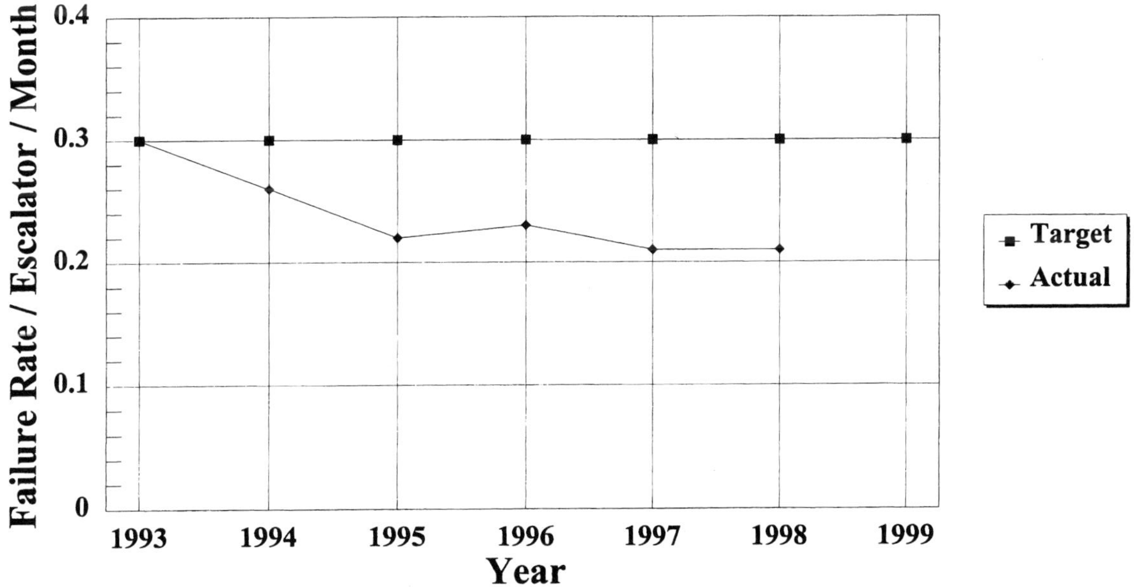

Staying on Track

C570/005/99

The development of an equitable cost attribution model for track deterioration

R J ARMITAGE and **D L COPE**
Scott Wilson Pavement Engineering Limited, Nottingham, UK
D CURTIS
Interfleet Technology Limited, Derby, UK

SUMMARY

This paper outlines a credible mathematical model for predicting the rate of settlement of any section of track in response to either current or proposed future traffic. The model provides a mechanism for equitably allocating the costs of track deterioration to its causation by particular traffic types. The development of such a model is of value to all members of the railway industry and will help to identify appropriate levels of expenditure for both track maintenance and renewal activities. The paper highlights those particular site investigation techniques and methods of analysis which have made the development of such a predictive model possible.

1. INTRODUCTION

The mechanism by which the Government supports financially the operation of train services in the United Kingdom is via the franchises which are allocated after competitive tendering to the Train Operating Companies. In turn, the Train Operating Companies purchase 'train paths' from Railtrack, the infrastructure owner, and Railtrack, in its turn, maintains and renews the infrastructure through competitively sourced contracts let primarily to infrastructure maintenance and renewal companies. Thus Government funding flows through the various railway organisations supporting the day to day operations and on-going development of the railway system in this country. The level of Government support is set to reduce by about 50% over the duration of the existing franchises, as competition and productivity improvements brought about by privatisation take effect.

However, at the present time there is little or no attempt made to attribute the costs of train operations to the need for maintenance and renewal of the infrastructure on a cost causation basis. Thus the cost of a train path is allocated rather arbitrarily based to some extent on its availability and the competition for such a path, but with little regard to the actual deterioration caused by the particular train operation over the particular route for which the train path has been allocated. This means that the access charges to the Train Operating Companies are largely fixed and there is little incentive for them to procure and operate trains which are more 'track friendly', ie. which impose lower forces on the track infrastructure, which in turn would lead to lower maintenance costs and extended life of the asset.

Equally the lack of an effective cost attribution process makes it difficult for either the infrastructure owner or the maintenance companies to forecast with any degree of certainty, the effect on the maintenance needs of the infrastructure of the introduction of trains operating at higher speed, higher axleloads, greater overall weight or with new characteristics such as tilting bodies.

Furthermore, the lack of a credible cost attribution mechanism makes it impossible to know with certainty whether the current and proposed Government funding is set at the correct level to maintain and develop the railway system to meet the needs of taxpayers in this country. Thus it can be seen that the lack of an adequate mechanism for attributing costs to actual operations undermines the whole basis of the present arrangements for funding railway operations in this country, and raises major questions of policy at the highest level.

2. THE PROBLEM

Why then has the development of an equitable cost attribution mechanism proved so elusive? On the one hand it could be argued that there is no need for such a model and the present arrangements suit all the parties involved. However, the authors believe that such a fundamental lack of understanding of the cause and effect of train operations is not a sustainable scenario. A firm understanding of the cost base is fundamental to the ongoing operation of any successful enterprise and without it the organisation cannot continue to operate efficiently and profitably.

On the other hand it is undoubtedly true that the development of a cost attribution mechanism that is fair and acceptable to all parties is a very difficult exercise and in the absence of such a model the parties may be quite happy to accept the arbitrary nature of cost allocation provided they do not perceive themselves to be disadvantaged by it.

If the above situation persists then it could be argued that there is little incentive to develop something better than the present arrangements. It may appear to be just too difficult. However the authors believe that it is possible to make some worthwhile progress in developing a cost attribution mechanism that the whole railway industry is comfortable with. To succeed it is recognised that the model must be able to be validated by reference to actual instances of cause and effect being demonstrated and supported by credible scientific evidence.

3. SOME KEY QUESTIONS

The successful development of the proposed model should ensure that answers to the following types of question can be satisfactorily provided:

<u>For the Infrastructure Owner:</u>

- What is the true cost of train operations on your infrastructure?
- Can rates of deterioration and the need for track renewals be accurately forecast?

- Can the effect of higher speeds, axleloads, train weights or tilting operations on your infrastructure maintenance costs be predicted?
- Is the proportion of damage caused between freight and passenger operations known?
- Are the costs charged by your maintenance contractors fully justified?

For Infrastructure Maintenance Companies:

- Can your maintenance activities be accurately planned?
- Can the effect of changing patterns of train operations on your future maintenance activities be ascertained?
- Is it known or understood why some sections of track do not respond to maintenance activities?
- Can you forecast when track at some locations will become life expired within the duration of your contract term?
- Can you be confident that you are applying the most appropriate maintenance techniques in all circumstances?

For Track Renewal Companies:

- Is there a clear understanding of what has caused particular sections of track to require renewal?
- Does the solution offered provide the most cost effective whole life cost solution to your customer, Railtrack, and can this be demonstrated?
- Do you know that the extent of your proposed renewal scheme will solve the problem?
- Are your renewal proposals optimised to minimise future possession needs?

For Train Operating Companies:

- Do your track access charges accurately reflect the cost to the infrastructure owner of your operations?
- What changes to track deterioration will your proposed future train operations bring?
- Could you make a case for reduced charges by operating and specifying more 'track friendly' rolling stock?
- Is the track geometry quality optimised to provide the desired level of comfort for your passengers?

For Rolling Stock Leasing Companies:

- Would you know how to specify more 'track friendly' vehicles in the future?
- Could you convince Train Operating Companies of the likely reduced track costs to enable them to re-negotiate track access charges?
- Can you do accurate 'what if ?' comparisons to determine the most cost effective future rolling stock to minimise track deterioration?

For Track Component Suppliers:

- Can you demonstrate that your current products are 'fit for purpose' in a changing train operating environment?
- Will they effectively support higher speeds, higher axleloads, greater tonnage or higher lateral forces from tilting trains?
- Do you know how your current products are performing in all circumstances?

4. THE SOLUTION

In order to develop a credible cost attribution model three components are required:

1) A vehicle dynamics simulation model
2) A track deterioration model
3) A mechanism for linking the two together

The output has to be structured in such a way that it can answer the questions posed above (and others); the output has to be credible and its validity demonstrated by reference to actual events.

A number of credible vehicle dynamics packages have been developed in recent times including VAMPIRE, MEDYNA, NUCARS and others. Studies using these models have been used to demonstrate their validity to the extent that they now form part of the vehicle design and acceptance process in many countries. Such programs also have the facility for modelling standard train formations and can therefore be used to build up a model of the operations on any particular route. One of these existing packages can form the basis for the vehicle dynamics simulation part of the above process without significant change.

Of much greater difficulty has been the process of developing a credible track deterioration model. A number of attempts have been made, particularly MARPAS developed by British Rail Research (now AEA Technology-Rail), by Professor Sato in Japan and more recently by the Swedish and South African railways. However so far none of the models is sufficiently comprehensive to allow their use for accurate predictions of deterioration of track under prescribed traffic flows, which is an essential requirement if site specific planning is to be undertaken. Only generic predictions have been possible so far which are valuable in some areas of strategic planning but do not answer the specific types of question posed above.

The breakthrough which the present approach represents is brought about by the ability to model for the first time the response of individual track sections as a result of developing a thorough understanding of the deterioration mechanisms, the relevance of individual track parameters to these deterioration processes, and the ability to productively collect relevant track data.

The mechanism for linking the two models is to use an iterative, stepwise simulation of the passage of a single wheelset, both statically and dynamically, over the model of the track structure, if necessary at every sleeper, and taking account of the initial track geometry and the relevant settlement algorithms pertinent to the actual track properties (Figure 1). Then to calculate the revised track geometry as a result of the inelastic settlement caused by the individual wheelset, prior to passage of subsequent wheelsets. Passage of further trains, each characterised by both dynamic and static properties, allows accumulation of the predicted settlement data and prediction of changes to other trackbed properties such as ballast condition. This enables the total deterioration under the actual traffic mix to be determined and attributed to each train pass. Decision making interventions may be prescribed and incorporated resulting in changes to track geometry details, or settlement and degradation of the track elements may be calculated to identify failure against prescribed modes.

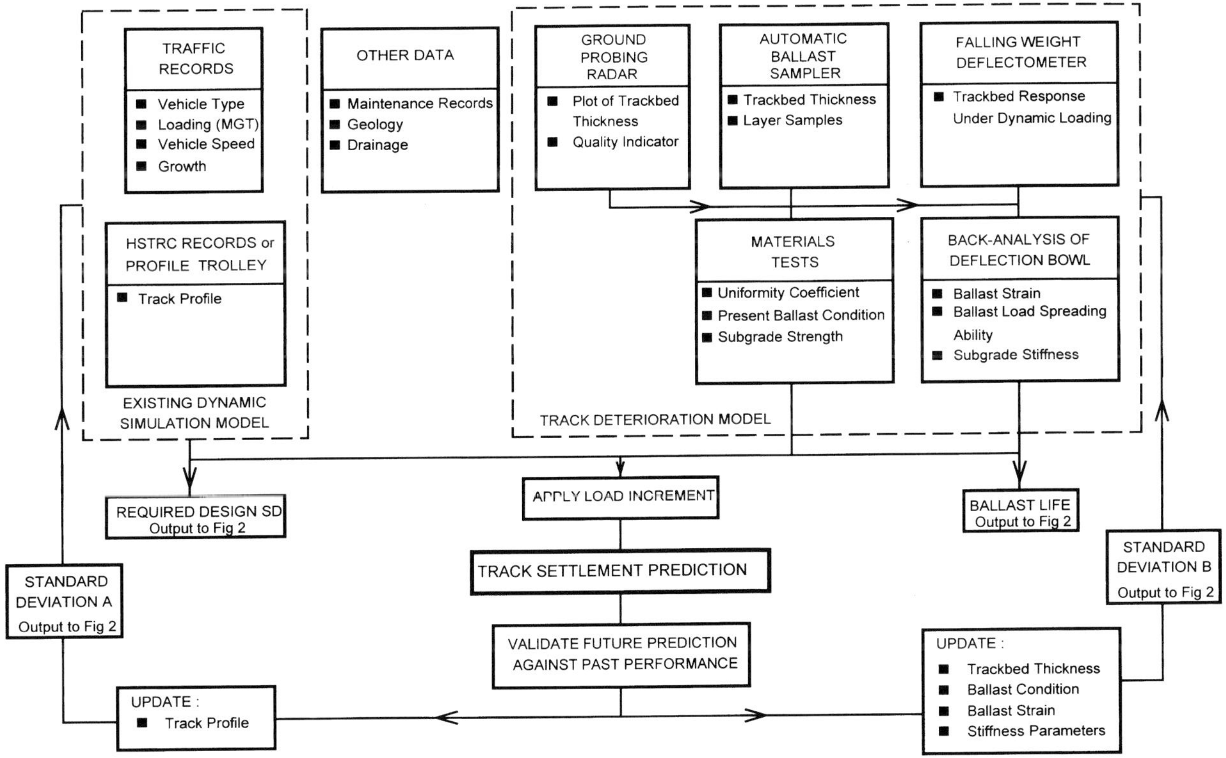

FIGURE 1 : OUTLINE OF TRACK SETTLEMENT PREDICTION MODEL

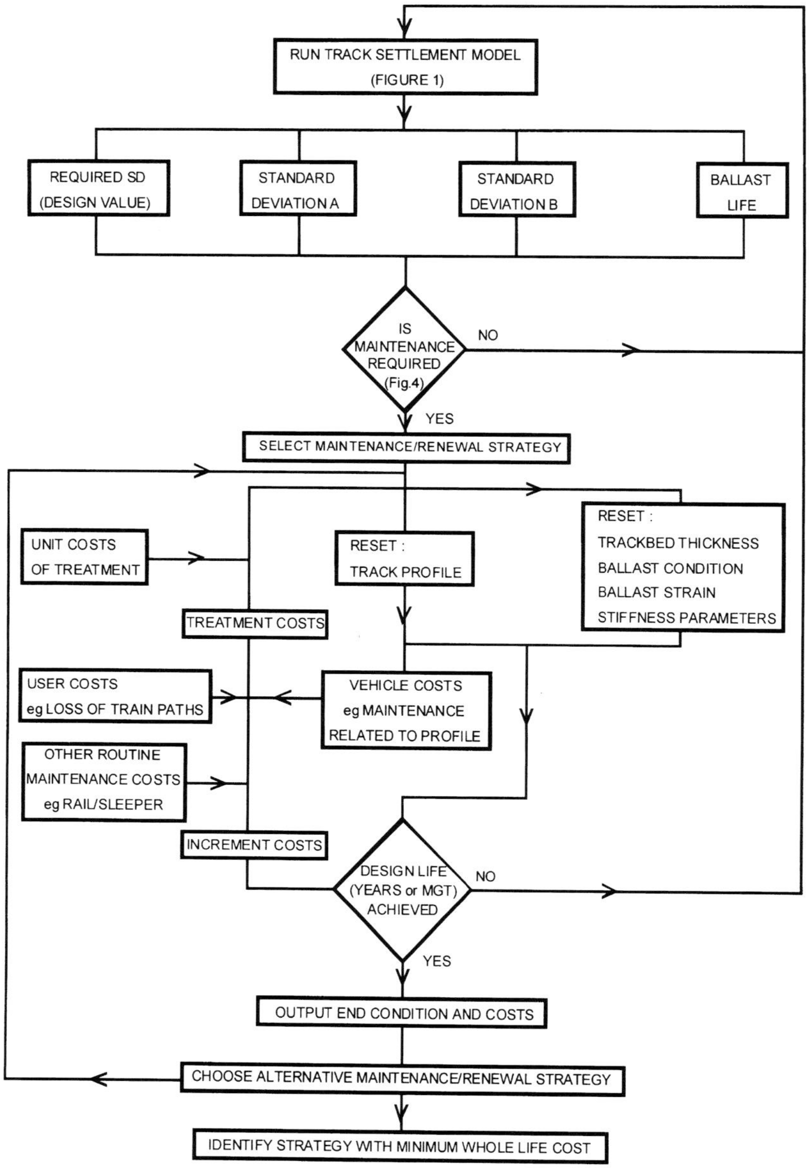

FIGURE 2 : MAINTENANCE OPTIMISATION USING COST ATTRIBUTION MODEL

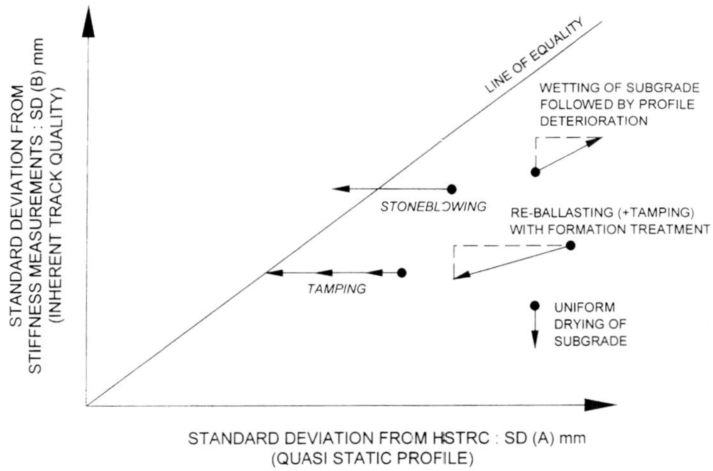

FIGURE 3 : IDEALISED STANDARD DEVIATION PATHS ON PLOT OF SD (A) VERSUS SD (B)

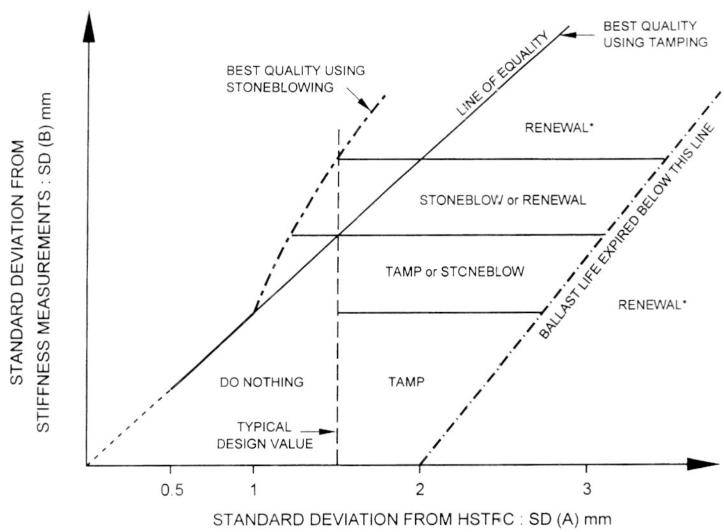

RENEWAL* = Combination of re-ballast/ballast clean with formation/drainage improvement followed by tamping, then stoneblowing, all as appriopriate. Treatments to be developed from interpretation of detailed site investigation data, and optimised using procedure in Figure 2.

FIGURE 4 : SELECTION OF MAINTENANCE TREATMENT USING TYPICAL DESIGN VALUE OF 1.5

Clearly such a process allows 'what if ?' simulations to be undertaken (where the variables may relate to either vehicles, track or both), and permits identification of optimised intervention activities and proposals for renewal solutions which provide a minimum whole life cost (Figure 2).

5. TRACK DATA ACQUISITION

A number of previous attempts have been made to forecast track deterioration rates by reference to historic or statistical analysis of track geometry data pertinent to particular sites. However such methods are unable to forecast the effect of changes inherent in the deterioration process or the effect of changes in the traffic characteristics.

It is the development of productive and practical methods of track substructure evaluation which has provided a capability for the development of credible models. The models predict the rate of deterioration of track geometry by linking the settlement of the track to the loading applied by railway vehicles.

Three investigative techniques are available, two of which have been extensively used for the evaluation of highways, airfield runways and other bound pavements and the other has been specifically developed for the sampling of railway trackbed materials.

- Ground Probing Radar (GPR) : This non-destructive technique uses electromagnetic radar waves to identify the presence of layer interfaces between the different materials comprising the trackbed construction. The depth of interface can be estimated knowing the travel time for the waves and approximate material dielectric constant. If materials have become contaminated (eg. slurried ballast), there is no clear return signal since the interface does not exist. Typically a radar reading is taken between each pair of sleepers, and all the results are combined to give an effectively continuous profile.

- Automatic Ballast Sampler (ABS) : This device is used to obtain 65mm diameter samples through the trackbed, typically to 1m depth. If required a smaller 50mm diameter tube can be driven to a greater depth through the hole formed in the ballast. The samples are sealed in a plastic tube, enabling them to be returned to the laboratory for logging, photographing and any materials tests deemed appropriate. The logs are particularly valuable in understanding deterioration modes and calibrating the GPR profiles.

- Falling Weight Deflectometer (FWD) : The FWD is a non-destructive device designed to apply a known impulse load, via a sleeper, to the trackbed. Transient deflections of sleeper and the ballast surface are measured to an accuracy of a few microns. The data can be interpreted to determine several indicators of ballast and subgrade performance, to provide valuable information on the behaviour of the trackbed under wheel loading. The quantitative evaluation of subgrade stiffness characteristics provides an essential element in the understanding of subsequent track settlement behaviour.

The use of these techniques combined with information derived from routine track geometry measurements made by the High Speed Track Recording Coach (HSTRC) or Track Recording Unit (TRU), or by measurements of the unloaded track top profile using trolleys; plus other readily available data such as rail type, sleeper type, sleeper spacing and rail pad characteristics; together with data on historical, present and future traffic loading, provide all the elements required to use the predictive model proposed.

6. TRACK SETTLEMENT PREDICTION

Figure 1 details the input data which is required. This consists of three elements which are normally readily available (Traffic Records; HSTRC Records; Other Data) and the remaining elements obtained from site investigations, materials tests in the laboratory, and analysis of data.

The track settlement prediction uses both the Dynamic Simulation Model and the Track Deterioration Model, and the output is used to update the original base data. Comparison of predictions with observed past performance helps to validate the model, for which HSTRC trends and records of past maintenance are particularly useful. The output is also used to calculate the profile standard deviation in two ways. Firstly, using the updated track profile, and secondly, by predicting the "inherent track quality" resulting from the variation in trackbed and subgrade stiffness (load spreading ability) derived from Falling Weight Deflectometer deflections.

Figure 3 shows an idealised plot of these two parameters, referred to as SD(A) and SD(B) respectively. On the figure a number of typical Standard Deviation (SD) paths are plotted. These show predictions of changes in SD which occur as a result of environmental changes to subgrade or maintenance activities on subgrade and trackbed. Research has shown that data from most sites lies on or below the line of equality. Tamping can improve the SD(A), but will have negligible effect on SD(B) since it does not change the inherent stiffness of the trackbed and subgrade. The line of equality represents the best quality that can be achieved by traditional maintenance activities, such as tamping of ballast. Further improvement (ie. movement downwards and towards the origin) is impossible unless the subgrade is improved at the same time.

Figure 4 provides another SD(A) versus SD(B) plot, but, instead of showing SD paths, areas on the diagram have been identified which represent the most suitable maintenance treatment. It should be noted that stoneblowing, although considered unsuitable for lowering the SD below 1.0, has been shown to achieve a lower SD(A) for poorer quality track than is possible with tamping. However, in the example where a typical design value of 1.5 is required, even stoneblowing will not be effective if the SD(B) value shows that there is inherently poor, variable and/or very soft subgrade.

It has already been shown (Figure 3) that tamping moves the SD path horizontally to the left. In contrast ballast deterioration (eg. due to fines generation/infiltration combined with moisture) will move the SD path to the right (and possibly slightly downwards as the load spreading ability deteriorates). Eventually the SD path enters the region where only renewal is appropriate. However, the optimisation of this activity ideally requires interpretation of all the detailed site investigation data, combined with understanding of the constraints (eg. access limitations) pertinent to the site. Furthermore, optimisation should consider the costs of the treatment and other predicted maintenance activities throughout the life of the trackbed. Ideally 'user costs' and 'vehicle operating costs' should also be included, although these aspects are currently outside the scope of the model. Figure 2 illustrates in more detail how the various components can be made to interact in order to develop a maintenance strategy optimised for minimum whole life cost.

7. MODEL DEVELOPMENT

A key element in the development of the model is the production of accurate settlement prediction algorithms based on the known or derived characteristics of the track subgrade materials. Clearly the accuracy of such algorithms is essential if the predicted settlement results are to be used with confidence and these rely on the collection of data from real sites.

A number of pilot sites have been subjected to the full range of investigations outlined above and these will be joined by further sites subjected primarily to investigation by ground probing radar only. This latter technique has the advantage of covering the largest amount of ground in the shortest time and can be used to highlight areas requiring a more detailed investigation.

The data from these investigations will provide sufficient information to develop generic algorithms which will represent the main combinations of trackbed parameters and trafficking variations across the Railtrack network. Development of the model to this stage will prove very useful to Railtrack in ensuring that particular routes that may be subject to major changes of traffic patterns are able to cope with the proposed changes.

It is impossible to gather data from all Railtrack infrastructure immediately and therefore a two stage implementation process is proposed. It is envisaged that the model will initially be used at Network level (or perhaps on major train corridors such as the West Cost Main Line). Following on from the successful demonstration of the model at the Network level, the algorithms will be particularised on a site by site basis at the so called Project level. In this application, use of all the main site investigation elements can be brought into play, particularly on those areas identified as having the least life. The model can be used to optimise renewal solutions from which productive maintenance schedules can be developed. In addition detailed settlement predictions can be made in relation to either existing or proposed new traffic flows. Future monitoring of settlements and trackbed condition will lead to improved algorithms.

The development of the model will provide benefits to all parts of the railway industry. Furthermore, the site data collected will not only be of value in developing the settlement algorithms, but has an inherent value in providing a more comprehensive approach to site investigation than traditional methods in advance of planning track renewals. The data from the pilot studies is already assisting Railtrack in this process.

8. ACKNOWLEDGEMENTS

The authors acknowledge the financial support of Railtrack and the valuable contributions being made to this project by their colleagues, in particular Dr P. Sharpe at Scott Wilson Pavement Engineering and Messrs R. Garner, I. Hyde and I. Hills at Interfleet Technology Ltd.

C570/011/99

Towards fault-free Permanent Way for MTR Corporation

R M KEEFE and **I C K LAI**
Mass Transit Railway Corporation, Hong Kong

SYNOPSIS

MTR Corporation operates one of the world's most densely utilised urban mass transit systems with high expectations for service reliability. Demands on the track and its management are high. The efforts made to continuously improve the performance of the track and reduce failures are examined, results analysed and conclusions drawn. Lessons learnt from management of the original three lines were input into the trackwork specification for construction and maintainability of the two most recently constructed lines, which opened in 1998.

1. INTRODUCTION

The MTR Corporation network now consists of 5 lines. The three Urban Lines comprising the Kwun Tong Line (KTL), Tsuen Wan Line (TWL) and Island Line (ISL) opened in stages between 1979 and 1989. These lines operate at a maximum speed of 80 km/h with headways as low as 105 seconds in the peak hours.

The two new Lantau Lines comprising the Tung Chung Line (TCL) and the Airport Express (AEL) opened in the middle of 1998. These operate at 135 km/h with ultimate service frequencies of 4.5 minutes, resulting in headways of 2.25 minutes on shared track. The construction and operation of these lines, colloquially known as the "Airport Railway", has been described in detail by Crighton and Budge-Reid (1).

The design and construction of MTR's initial system, which opened in 1979, has been the subject of various technical papers such as those by Sharpe et al (2) and Taylor et al (3). Apart from the change of rail type from BS11-90A used on the KTL and TWL to UIC60 for the ISL and a short extension to the KTL, the basic track design remained unchanged.

With over 10 years experience, an opportunity was taken in 1992 to review the specifications currently used for the Urban Lines. The objective was to make a step change in the reliability to drive towards a fault free Permanent Way for the new Lantau Lines.

2. MTR's OPERATING ENVIRONMENT

MTR operates on commercial principles within a competitive free market with many alternative parallel modes of transport such as buses, mini buses, taxis, ferries and trams.

MTR's mission is

- to be the world's most customer oriented railway

It is required by its Ordinance to exist

- for the principle purpose of constructing and operating, on prudent commercial principles, a mass transit railway system having regard to the reasonable requirements of the public transport system of Hong Kong

The Corporation carries out regular customer surveys, which show that of all the attributes required by our passengers, a speedy and reliable arrival is the core determinant of success, or failure. This may be summed up in Figure 1 below, using Kotler's "Augmented, Tangible and Core Product Concepts" (4).

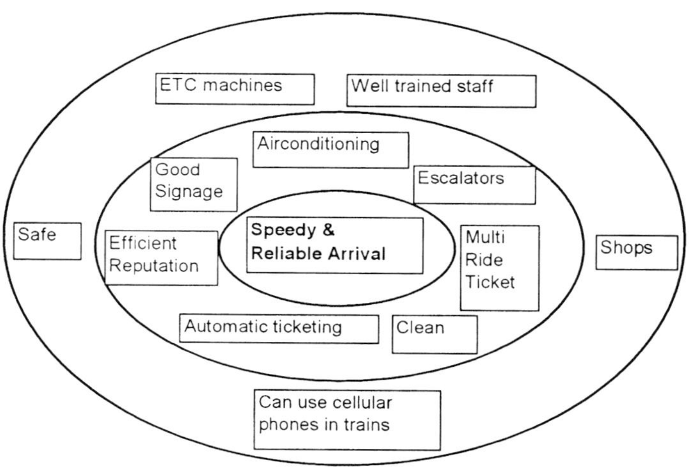

Figure 1 – Augmented, tangible and core product for an MTR journey

With a daily patronage of around 2.3 million on the 43.2 route km Urban Lines, there is a high density of traffic, with capacity in the peak period for up to 85,000 on one line in one direction. Any disruption whether caused by the infrastructure, rolling stock, operational procedures or the passengers themselves can quickly result in severe overloading and disruption to other transport modes.

Effective management of the infrastructure to avoid failures which do, or have the potential to, cause service delays is required to ensure that the overall service delivery to the customer meets their requirements. Passengers demand increasing customer service while the Corporation seeks continuous improvement in cost effectiveness. One method of achieving both is to reduce the number of faults. This is as true of the Permanent Way as it is of the other engineering disciplines.

Permanent Way faults need to be prevented so that the number of incidents impacting on the passenger service it is reduced, ideally towards zero. These may be divided into:

- failures which actually impact on service
- failures which would have impacted on service had the circumstances been different

In identifying the opportunities to reduce and eliminate faults, the people (ie. the staff), the process and the products (or components) have to be evaluated and necessary changes made.

Analysis of past failures and thorough testing of new processes or products is necessary to ensure potential for future faults is reduced. Staff need to be trained, not only in the correct application of the processes through training, but also in recovery techniques.

Whilst track regular maintenance is necessary to ensure the system remains operational, each time maintenance or renewal is undertaken there is the possibility that incorrect application of the process will lead to a fault. Therefore, reducing the number of times intervention is required will have the potential to reduce staff initiated failures.

3. REDUCING FAULTS ON THE URBAN LINES

The urban lines feature track using a mixture of BS11-90A rail (67 track km) and UIC60 rail (33 track km) predominantly supported on a continuous reinforced concrete plinth and secured using the Pandrol rail fastening system.

With a continuous rail plinth, a 1500V DC overhead line traction supply system and tunnels constructed adjacent to and underneath the harbour of Hong Kong, rail to earth insulation was and remains difficult to achieve, resulting in rail foot corrosion.

The continuous plinth has a high track stiffness compared to ballasted track. This was exacerbated by poor performance of the continuous rail pad which contributed towards high corrugation growth rates.

With two rail sections of differing gauge corner profiles, poor compatibility of the wheel and rail profiles was achieved. Over stressing of the gauge corner resulted and together with a lubricant, which too easily migrated on to the rail head, endemic gauge corner shelling resulted.

These problems were discussed by Dring and Bygate (5) and a programme commenced to remedy the root causes by the development and application of a rail management model described by Keefe and Soeleiman (6). This involved the development and application of modified wheel and rail profiles, use of a new lubricant, an improved continuous rail pad, grinding cycles based on individual curve characteristics and latterly the introduction of heat treated rail on sharper curves.

The application of the technical and managerial model have resulted in reduced rates of corrugation growth, reduced rail renewal due to gauge corner shelling and reduced rail renewal due to sidewear as described by Keefe and Fung (7). This has had the benefit that, as the rails require renewal less frequently, there is less chance that there will be a possession over run during renewal.

Although the application of a rail management model was very successful in reducing the number of faults from plain track, indicators showed that in the early 1990's the majority of failures resulting in service delays resulted from faults in turnouts and rail joints.

The predominant causes of failures resulting in delays were:

- Broken rails – mainly switch rails, rail expansion joints (REJ)or corroded plain rails
- Points failure due to broken stretcher bars or bracket
- Broken Huck / jaw bolts jamming the points
- Incorrect maintenance or damage to components by maintenance work

Other components which exhibited unreliability and hence the potential for service delay should a more serious failure occur included:

- Horizontal split web flashbutt welds
- Rail web cracks from welded traction current return connections
- Shelling of austentic manganese steel (AMS) crossing noses
- Bolt hole cracks in AMS crossings

As a result of experience from the initial operation of the KTL and TWL, it was realised that the BS11-90A rail had marginal spare capacity. The original rail section with its continuous support was selected, as described by Sharpe et. al. (3), by the contractor under a design and construct contract to minimise costs of installation and maintenance.

With the continually increasing traffic the decision was taken for the construction if ISL to use the larger UIC60 rail for that and future extensions. This was a far sighted decision as, although the maximum axle load is 20t, the annual tonnage of traffic carried on one track in one direction on parts of the network now reaches 46 million gross tonnes (Mgt.) per annum. Unfortunately these loads are carried on the part of the network which uses BS11-90A rail.

By adopting a more robust rail section it was hoped that the number of Permanent Way faults could be reduced.

Statistics from the Urban Lines (Figure 2) for 1992 – 1998 show that rail defects per track km are more than 5 times more likely to occur in BS90A rails than UIC60 rails.

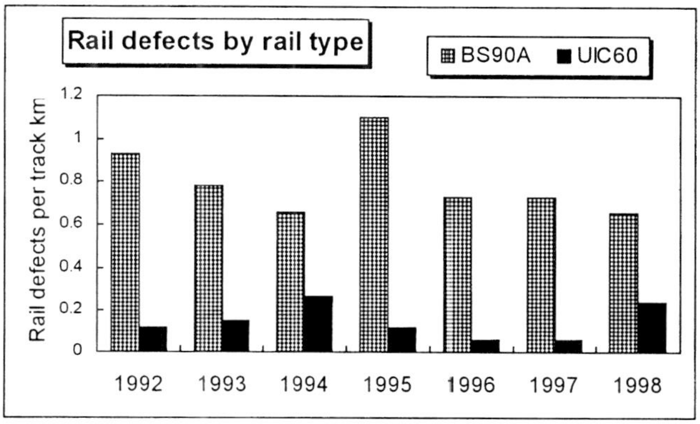

Figure 2 – Rail defects by rail type

With the introduction of UIC60 rail on the ISL, which opened in 1986, the opportunity was taken to introduce shallow depth asymmetric switches of the Zu1-60 profile. These together with the MJ80 point machine and VCC clamplock have resulted in a much lower rate of points failure and rail failures in UIC60 track than BS11-90A track.

In order to reduce the failures associated with BS11-90A trackwork, the decision was taken to upgrade BS11-90A turnouts towards the UIC60 standard. Direct replacement with UIC60 rail is not cost effective due to the fact that the majority of turnouts are directly fastened to the concrete slab in tunnels, so a Zu2-49 asymmetric switch rail was chosen to match the BS11-90A switch.

A broad range of solutions for the main track failure modes has been implemented over the period since 1992 as shown in Table 1 overleaf.

Table 1 – Solutions implemented on the Urban Lines to reduce faults

Fault / failure	Solution(s)	Notes
Broken switch rails	Introduction of Zu2-49 asymmetric switch to replace BS11-90A. Improved surface finish on machined surfaces to reduce stress raisers	Broken switches have only ever occurred in full depth BS11-90A rail section. All UIC60 turnouts use Zu1-60 asymmetric switches
Broken rail expansion joints (REJ)	Adoption of Vo1-49 thick web REJ blades	Broken REJs have only ever occurred in BS11-90A
Broken corroded plain rails	Improved leak sealing of tunnels Use of improved continuous rail pad Improved inspection and replacement of corroded rails	Due to the trackform design the desired rail to earth insulation level is difficult to achieve with a continuously supported rail
Horizontal split web flashbutt welds	Peening of flashbutt welds in high leg of sharp curves to reduce residual stresses	Peening now replaced by heat blocks for better process control. Also can be applied in the field
Broken stretcher bars / brackets	Use of low friction slides to reduce variability in driving forces	Requirement for lubrication eliminated
Broken Huck / jaw bolts jamming points	Replacement of mis-aligned bolts Zu2-49 asymmetric switch eliminates need for Huck / jaw bolts	
Gauge corner shelling, high corrugation growth rates and sidewear	Improved wheel / rail profiles with revised grinding cycles Improved continuous rail pad and rail lubricant	Developed and applied with a holistic approach using a technical / economic rail management model
Incorrect application of process or damage to components during maintenance work	Education and training of staff Investigate causes of human error Clearer job instructions and standards	
Rail web cracks from welded traction current return bonds	Use of a brazed connection	Now being replaced by a bolted connection
AMS crossing nose shelling	Preventative grinding of crossing noses	
Fishbolt hole cracks at AMS crossings	Cold expansion of AMS crossing fishbolt holes Trial of AMS crossing with flashbutt welded legs	

Despite the tonnage of traffic having increased by 10% since 1991 (Figure 3 overleaf), the average age of defective rails has increased by 90% over the same period (Figure 4 overleaf).

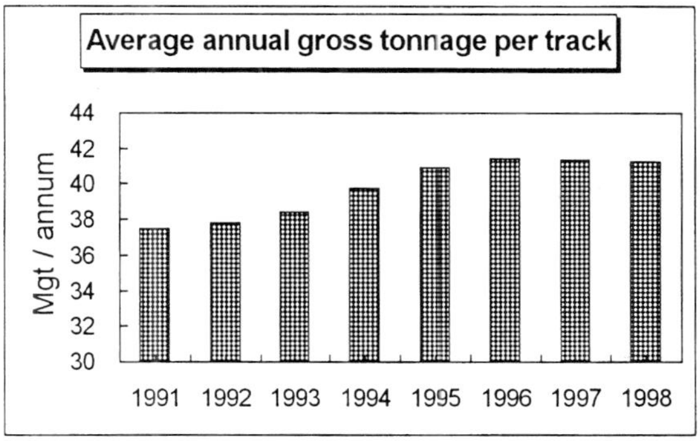

Figure 3 – Average annual gross tonnage carried per track on Urban Lines

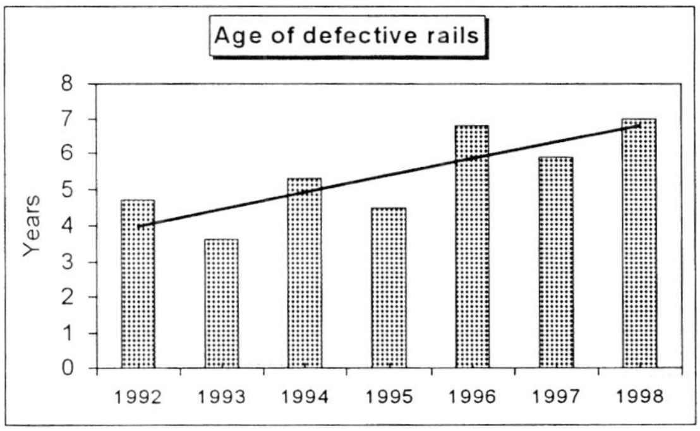

Figure 4 – Average age of rail at time defect discovered

A rail defect is one which requires the rail to be prematurely removed from track. Most dramatically the number of broken rails per million train miles over the last 10 years has decreased, with no broken rails being recorded in 1997 or 1998 (Figure 5 below).

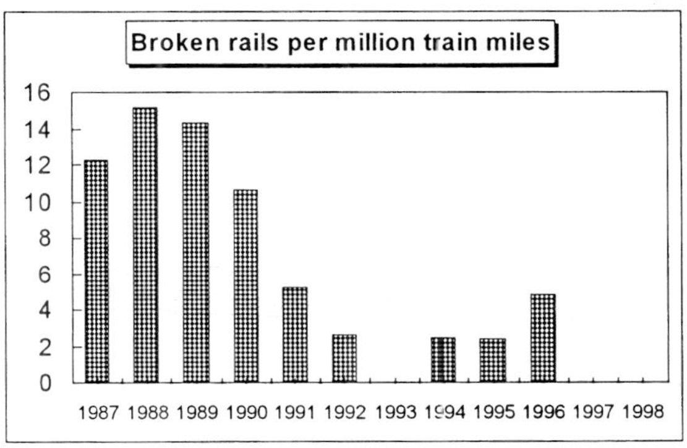

Figure 5 – Broken rails per million train miles

It is evident that the strategy has resulted in far fewer Permanent Way failures

Figure 6 however shows that this has yet to be translated onto measurable improvements in the number of delays to service attributable to the Permanent Way since 1994, although there has been a steady improvement since 1996.

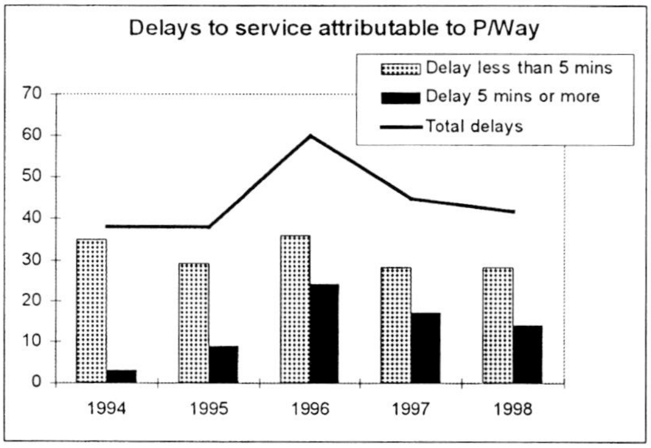

Figure 6 – Delays to service attributable to Permanent Way

At the end of 1994, the brazed traction current return connection was introduced. By 1996 an average of 1 block was detaching per day resulting in an increase in delays to service due to track circuit failures. Staff formed a work improvement team to identify the source of the faults and increase the reliability of the connection.

If the downward trend from 1996 was to be continued it was considered necessary to further examine the relationships between the failure and the delays that ensued to:

- Reduce the total number of delays
- Reduce the effect of the delays on passengers

Given that MTR operates a very intensive service, the impact on train service of a delay is also affected by the knock on effect of the delay on the initial train on the subsequent trains behind, until such time as normal service can be resumed.

Delays to service are measured as "initial" and accumulative". When preparing statistics it is the worse of the initial or accumulative which is counted.

An initial delay is the delay to the first train involved in the incident.

An accumulative delay is the delay to the train worst affected compared to the timetable.

In order to minimise the effect on passengers it is desirable that if an initial delay occurs, that the accumulative delay is minimised. This is a function of:

- headway, as a shorter headway reduces recovery time
- ability to operate in degraded mode
- recovery techniques adopted

Figure 7 below shows the analysis of initial versus accumulative delay for all delays attributable to the Permanent Way where the accumulative delay was 5 minutes or more.

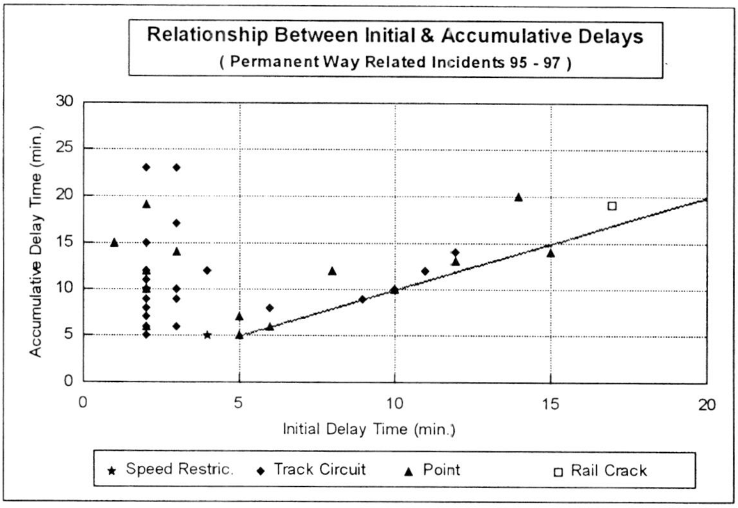

Figure 7 – Relationship between initial and accumulative delay

It can be seen that when the initial delay is greater than 5 minutes the accumulative delay is not usually much longer. However for delays of less than 5 minutes there is no apparent relationship between initial and accumulative delay time.

Whilst the objective of the Permanent Way section is to strive towards providing a fault free system and therefore not causing delays at all, it is clear therefore that if faults occur that the manner in which they are handled has a great impact on the passenger. The engineers and the operators have to work together to understand how the effect of delays, once they occur, can be minimised.

Figure 8 overleaf shows the analysis of the time of day at which the delay occurred for the delays shown in Figure 7.

It is evident that the majority of faults resulting in delays are found when trains first run in the morning. Reducing the number of faults left or caused by maintenance activities is therefore a priority.

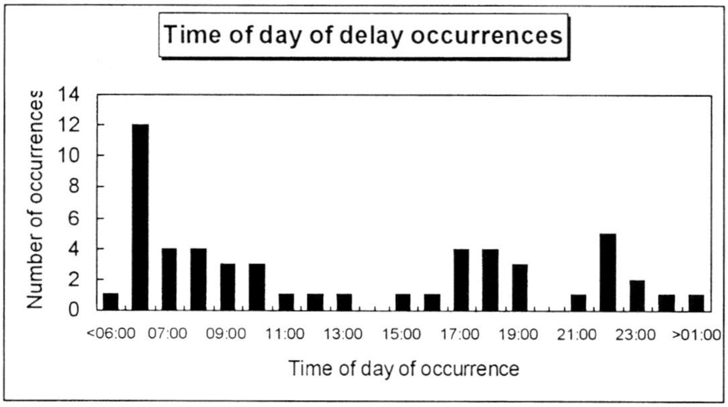

Figure 8 – Time of day of delay occurrence

3. APPLYING THE LESSONS LEARNED TO THE LANTAU LINES

A major extension to the MTR network, the Lantau Lines, as described by Crighton and Budge-Reid (1) gave the opportunity to start with a clean sheet of paper as far as the track design was concerned.

The Airport Express is a premium service designed to offer a speedy and reliable journey to and from the new Hong Kong International Airport and aimed at both arriving and departing passengers. In addition it features in town check in (ITCI) service at both Hong Kong and Kowloon stations.

Passengers luggage, once checked in, is carried in sealed baggage containers on the airport Express trains. Given that the ITCI has a cut off of only 90 minutes before scheduled flight departure, it is critical that the Airport Express service is reliable both for the passengers and their checked-in luggage.

The primary concern of the Permanent Way engineers was to economically provide a high reliability Permanent Way for train speeds of 135 km/h.

The first stage was a joint review by the Project designers and the Operational Railway managers of the then current specifications for trackwork, which were based on the ISL standard of 1983 for an 80 km/h railway and in-service experience as described previously in this paper.

As a number of problems had been identified with the wheel rail interface (3), (6), (7) it was concluded it was necessary to appoint a specialist consultant, BHP Research Melbourne Laboratories to review the most appropriate wheel rail interface conditions. These could then be included in both the Permanent Way and Rolling Stock specifications.

As the rolling stock on the LAL is captive, there was no need to consider inter-working of rolling stock, other than engineers' trains from the Urban Lines.

It was concluded that standard profile UIC60 rail with a P8 wheel profile was most appropriate. The rails were selected to be Grade 900A MHT (350 BHN) on curves of 1000m or sharper. Above this Grade 900A (285 BHN) rail would be used. Rail lubrication was recommended for curves of 700m or less.

The review of the ISL specification concluded that the majority of the systematic failures could be addressed in the technical specification for the LAL trackwork:

- stray current corrosion
  - very low permeability tunnel linings were specified
  - discrete support using the LVT system would improve rail to earth insulation. 250 ohm km was specified on non-ballasted track compared to 60 ohm km specified for the ISL. On the Urban Lines readings of less than 15 ohm km actually measured on the continuously supported track.
- rail corrugations and shelling
  - apply results of wheel / rail interface study
  - track support stiffness on non-ballasted track comparable with ballasted track by use of the LVT system
- rail failures
  - fully welded track layouts
  - asymmetric switches
  - peening of flashbutt welds in the high legs of sharp curves
  - use of brazed traction current return connection. In 1992/3 when the specifications were being prepared it was not predicted that subsequent reliability problems would be found with the brazed connection. At that time the laboratory testing showed the connection was reliable.

Despite an aggressive construction programme of just 43 months the HK$35.1bn (US$4.5 bn) railway opened on time with the first TCL service operating on 22 June 1998. The first passenger carrying Airport Express service operated coincident with the opening of the new Hong Kong International Airport on 6 July 1998.

After 10 months operation, the LAL and in particular the Airport Express have been critically acclaimed for the level of service provided and ease of use for passengers with luggage. The Airport Express alone is now carrying around 900,000 passengers per month.

In those 10 months there has only been one delay to service attributable to the Permanent Way, compared to 34 in the same period on the Urban Lines.

4. CONCLUSION

It has been identified that the original design of the Permanent Way on the MTR was one which resulted in a number of systematic faults once the demands on the Urban Line system increased close to its design capacity.

Through careful identification and analysis of the faults, measures have been implemented to reduce failures on the Urban Lines. The evidence suggests that the underlying causes of failures have been addressed and the number of faults has been reduced.

The construction of the Lantau Lines gave the Permanent Way engineers an opportunity to review the shortcomings of the Urban Lines and specify a Permanent Way which had a much lower propensity for faults to develop.

Initial operation results on the Lantau Lines show this has been successful in that the faults resulting in a delay to train services is an order of magnitude better than the Urban Lines.

The Lantau Lines specification is now being adopted as standard for all future extensions to the MTR network.

REFERENCES

(1) Crighton G and Budge-Reid A; Hong Kong International Airport railway. Proceedings, Institution of Civil Engineers Civil Engineering Special Issue, November 1998.

(2) Taylor R, Taylor M, Greig W and Couperthwaite R; Hong Kong Mass Transit Railway Modified Initial System: design and construction of above ground works and trackwork. Proceedings, Institution of Civil Engineers, Part 1, 1980.

(3) Sharpe D, Dring J and Singal B; The design, operation and maintenance of a mass transit system. Track technology for the next decade, Thomas Telford, London

(4) Kotler P, Principles of marketing, 6th ed.; Prentice Hall, 1994.

(5) Dring J and Bygate J, Hong Kong Mass Transit Railway concrete track support system. International Rail Track Conference, Auckland, New Zealand, 1989

(6) Keefe R and Soeleiman S, The development and application of a technical / economic rail management model to Hong Kong's Mass Transit Railway. Rail Track Association Australia Conference, Brisbane, Australia, September 1994

(7) Keefe R and Fung W, Recent developments in planning aids and track designs for the permanent way in Hong Kong MTRC. Exporail (Asia) 97 Conference, Kuala Lumpur, Malaysia, May 1997.

C570/015/99

Assessing railway vehicle derailment potential using neural networks

S IWNICKI and **J STOW**
Rail Technology Unit, Manchester Metropolitan University, UK
H J PARKINSON
PB Kennedy and Donkin Limited, Manchester, UK

ABSTRACT

Current methods for ensuring the safe running of railway vehicles assess the track and vehicle condition against fixed limits. Any exceedence of these limits requires remedial action to be taken. The setting of these limits is based on past experience or on computer modelling of vehicle track interaction. This paper describes the initial results of a novel method aimed at predicting vehicle behaviour from track measurements using an artificial neural network. The speed of the neural network method would allow quick analysis of all the data measured by the track recording coach and would also allow maintenance decisions to be based on the effect of track condition on the vehicle behaviour rather than on simple limits.

1. INTRODUCTION

Track recording vehicles regularly measure the vertical and lateral displacement of each rail , amongst other variables. Warnings are given when certain levels of irregularity are exceeded and simple algorithms are used to detect scenarios such as "cyclic top", which is known to contribute significantly to derailment risk. Thresholds are set on the basis of past experience but to ensure safety this can often result in a high number of false alarms. Additionally there is potentially a huge volume of data provided by the track recording vehicles which is difficult and labour intensive to analyse manually. Specialised computer modelling using packages such as VAMPIRE or MEDYNA can be used to predict the response of a vehicle to the particular track and to calculate the derailment ratio but this is time consuming and cannot realistically be carried out for all measured data and for all vehicle types.

The main input to the model of a rail vehicle is the rail alignment in the vertical and lateral directions and the variation in cross level and possibly gauge. This input is made up of random, periodic or deterministic components. Periodic components depend on rail section lengths between fishplated joints (if present), sleeper spacing, ground undulations and other regular features of the track. Random and discrete variations also exist in these inputs, for example track roughness and damaged rails. The tangential forces between the wheel and rail are calculated according to the methods developed by Kalker [1] and require a description of the wheel and rail cross sectional geometries. The likelihood of derailment is affected by the vehicle suspension as well as the track condition and assessment of derailment potential is not simple. Assessment is normally based on the derailment quotient (lateral/vertical force) at each wheel or on the level of wheel unloading.

The speed of the Neural Network NN method would allow all measured track to be effectively filtered and track likely to cause derailment of a particular vehicle be immediately flagged at the time of measurement The intelligence of the system allows complex interactions to be included rather than the simple limits currently used. This will allow better direction of the available track and vehicle maintenance efforts and should lead to safer running. Indications can also be given for likely levels of passenger comfort when a vehicle runs over the section of track.

To validate the method actual derailment cases need to be used to assess how accurately the NNs would have predicted the occurrence.

2. THE NEURAL NETWORK

A neural network is an interconnected assembly of simple processing elements, units or nodes, whose functionality is loosely based upon the human neuron. The processing ability of the network is stored in the inter-element connection strengths or weights, obtained by a process of adaptation to, or learning from, a set of training data. NNs are trained by inputting variables that correspond to known outputs. The network is initialised for training by assigning random weightings to the interconnections. If the NN gives the expected output, the weights of the connections leading to that output are strengthened. If an unexpected output is given, the weights are weakened. In this way the network can be taught to predict the output which will be the response to a new set of input data. A simple neural network consisting of 1 layer of 4 neurons each with 3 inputs is shown in figure 1.

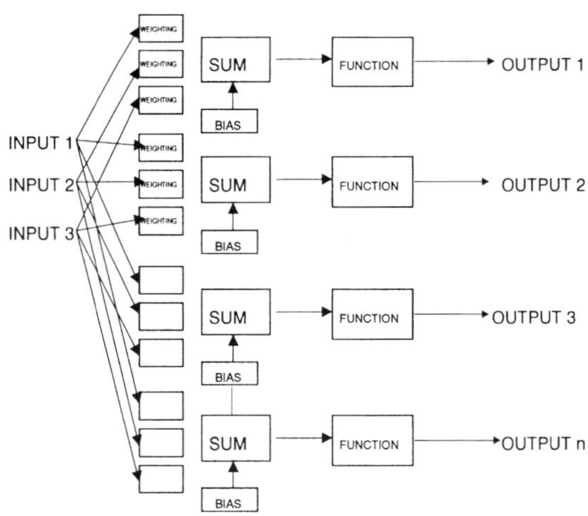

Figure 1. A layer of neurons

The basic neuron consists of a set of inputs each multiplied by a weighting factor before being summed and passed to the output via a simple transfer function. This function can be a go/no-go switch, a linear multiplier or any other suitable function. There is also an offset or bias available that can be applied to the sum of the weighted inputs. The neurons are assembled into layers and the sets of weighting and bias values govern the way that the output values relate to the input values for the whole layer.

The proposed intelligent track monitoring system is shown in figure 2.

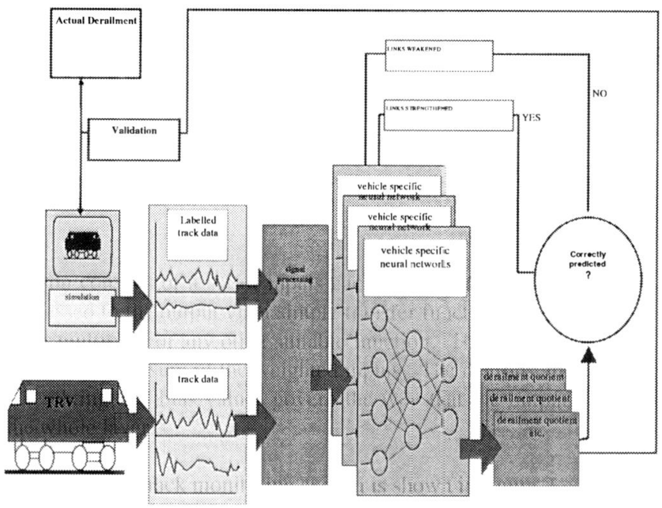

Figure 2. The proposed intelligent track monitoring system

Neural networks have been used in a wide variety of applications where complex systems need to be analysed and where a mathematical modelling method or a rule based analysis is not possible or is difficult to set up. Examples are pattern recognition or identification of abnormalities in biological or engineering systems. In the railway field NNs have been used to identify faults or to assist in optimising design [3]-[5]. Horwitz and El-Sibaie [2] proved that a neural network could match the prediction by a simple mathematical model of the dynamic vertical wheel load of a railway vehicle.

In this work the computer package MATLAB has been used to set up a suitable neural network. MATLAB offers a variety of basic neuron elements which can be built up into a network and then trained using data in files corresponding to the input and the required output.

3. TRAINING THE NEURAL NETWORK

Matlab offers an automatic training process where the input file is presented to the network and the weighting and bias values for the are optimised so that the required output is produced. In this work the input is real track recording coach data and the outputs are the wheel rail forces and the derailment quotient at each wheel calculated by a MEDYNA model of a Mk3 passenger coach with BT10 bogies. All data files are normalised before submission to the Network and rescaled after the prediction. Training takes place over a number of cycles or epochs and the number of epochs required to find the optimum values depends on the size of the NN and the acceptable level of the error.

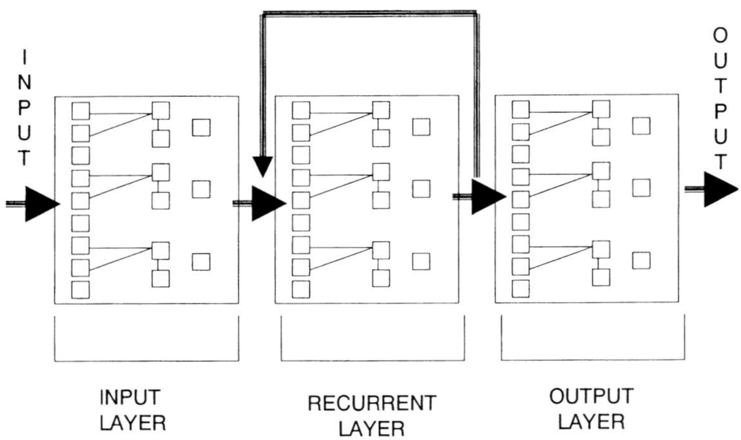

Figure 3. Recurrence in a neural network

The size of the Neural Network and the arrangement of the links between the neurons is important as it governs the way that the input and output relationships can be formed. In this work the input and output are time series and the output needs to be influenced by the previous values of the input and not just the current value. The Elman network provided by MATLAB was chosen for this work. It has a hidden layer with feedback (or recurrence) from the output. The hidden layer has transfer functions of the tansig type and an output layer with linear transfer functions was also used. Figure 3. shows the general layout of a neural network with recurrence.

4. THE MEDYNA MODEL

The training data for the neural network was prepared by running a simulation of a typical passenger coach on measured track. The vehicle chosen was a Mk3 coach with BT10 bogies with primary coil springs and hydraulic dampers on swing arms at the primary suspension and air springs at the secondary suspension. The model included representation of the primary and secondary suspension and the simulation package calculated the non-linear wheel-rail forces at

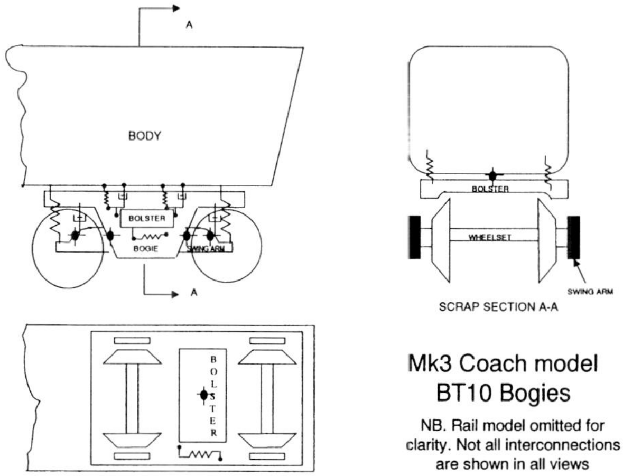

Figure 4 Medyna model of Mk3 Coach

each time step based on a description of the cross sectional geometry of the wheel and rail and using the contact program FASTSIM.

The simulation took about 10 hours and data files including the vertical and lateral rail displacements at each wheel and the outputs such as the vertical and lateral wheel-rail forces and their ratio and the acceleration in the main body of the vehicle were prepared after the run.

The vehicle model is shown in figure 4.

5. RESULTS

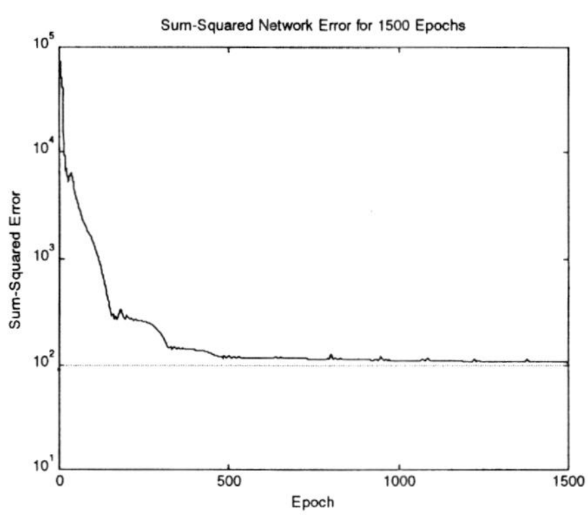

In practice the amount of feedback provided by the Elman network was found not to be sufficient to give good results. In recognition of the fact that the vehicle behaviour is very much influenced by the section of track that has just been passed over, it was decided to use a number of the previous data points from the track input file as inputs to the NN. This dramatically improved the ability of the network to match the required output. The reduction in the error as a typical training session progresses is shown in figure 5.

The NN took about 500 epochs to train and this training took about 1 hour on a 200 MHz Pentium processor. Running the network once trained is very fast with each 125m section of track taking less than 1s of processor time. This would mean that the output could be predicted at any reasonable measuring speed of a track recording coach.

Once the NN has been trained and gives satisfactory results with the training data unseen data is used as a test of the effectiveness of the network. The results presented below show that for a NN with 80 neurons in the recursive layer and with 120 preview points the prediction of the vehicle behaviour on track not previously seen by the NN is impressive. Vehicle speed is $20ms^{-1}$.

The plots show the load and derailment quotient predicted by the NN and the equivalent values from the MEDYNA simulation against distance along the track. The predictions of vertical load are generally within 10% but the lateral forces seem less easy to predict with large errors in some peak values.

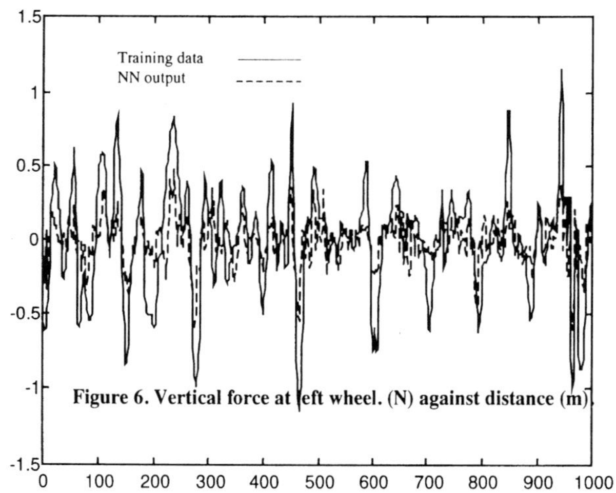

Figure 6. normalised vertical force at left wheel against distance (m). Training data and NN output (40 recurrent neurons and 80 preview points)

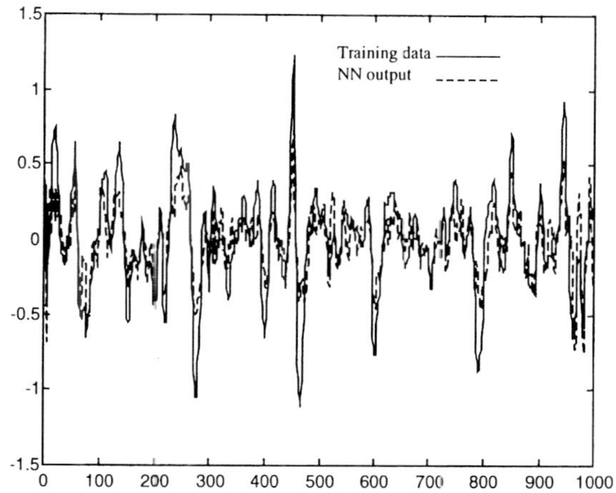

**Figure 7. Normalised vertical force at left wheel against distance (m)
Training data and NN output (80neurons and 80 preview points)**

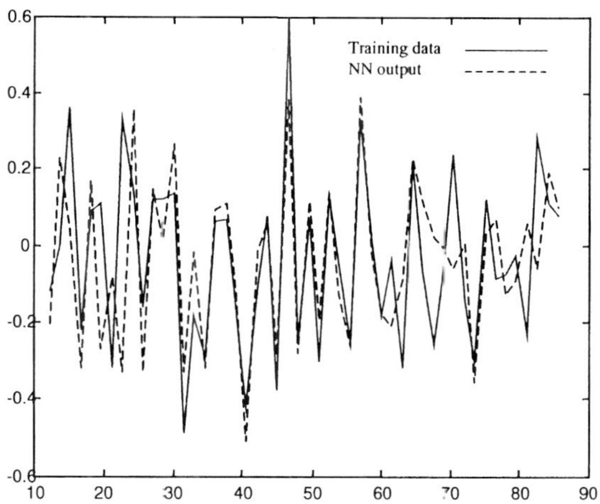

**Figure 8. Normalised vertical force at left wheel against distance (m)
Training data and NN output (80neurons and 120 preview points)**

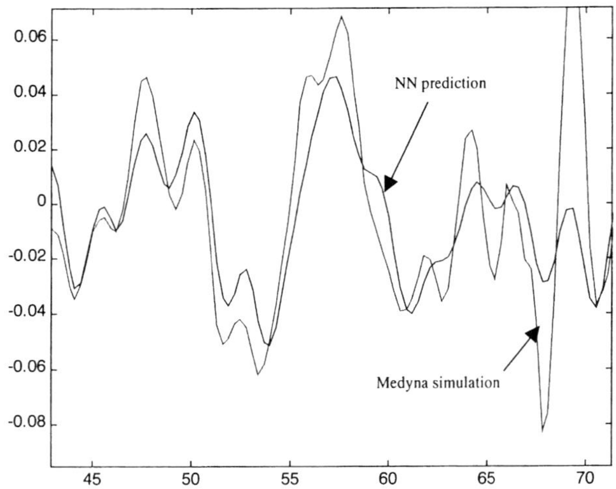

Figure 9. Lateral/Vertical force ratio at left wheel against distance (m)

CONCLUSIONS

This preliminary study has shown that is is possible to use an artificial neural network to predict the response of a railway vehicle to measured track input. The prediction of vertical force made by a neural network from track data comes very close to that of a computer simulation using a detailed model of the vehicle suspension and non-linear wheel rail contact model. The prediction of the lateral force and therefore the important lateral/vertical force ratio is less good and further work on the network for the lateral prediction is required.

The speed of the Neural network method should make it possible to filter a large quantity of measured track data and to focus maintenance effort on sections of track likely to excite dangerous or uncomfortable behaviour in vehicles.

ACKNOWLEDGEMENTS :

The authors would like to thank Railtrack and Serco Railtest for providing the measured track data.

REFERENCES

[1] Kalker J.J. 'The computation of three dimensional rolling contact with dry friction' Int. J. for numerical methods in Engineering Vol 14, 1979.

[2] Horwitz D. and El-Sibaie M., 'Applying Neural Nets To Railway Engineering', AI Expert January 1995

[3] Chow T.W.S. and Shuai O. 'Feedforward Neural Networks Based Input-Output Models for Railway Carriage System Identification', Neural Processing Letters 5: 127-137, 1997

[4] Choromanski W. 'Application of Neural Network for Intelligent Wheelset and Railway Vehicle Suspension Designs', Vehicle System Dynamics Supplement 25(1996), pp 87-98

[5] Gajdar T., Rudas I. and Suda Y. 'Neural Network based Estimation of Friction Coefficient of Wheel and Rail' IEEE 1997

Trains on Track:
Their Role in the Performance Matrix

C570/021/99

Implementing a train protection and warning system

R M KAUL
Halcrow Transmark Limited, Derby, UK

Synopsis

In July 1999 the Health and Safety Executive (HSE) issued 'The Railway Safety Regulations 1999'. These Regulations mandate the introduction of a train protection system that will cause the brakes of a train to apply automatically if the train passes a certain type of signal at danger or travels at excessive speed on a relevant approach. The deadline for the introduction of the system is 1st January 2004.

The Regulations effectively mandate the implementation of the Train Protection and Warning System (TPWS) that has been developed by Railtrack. This system consists of both trackside and trainborne equipment that interface to provide 'Trainstop' and 'Overspeed' functions.

This paper will introduce the principal features of TPWS and the function of, and the interface between, the trackside and trainborne equipment. The main body of the paper will describe the challenge of implementing the trainborne TPWS equipment within the mandated timescales, in a manner that ensures a positive contribution towards the fault free interface of the train with the infrastructure.

1 Background

In January 1994 both the British Railways and Railtrack boards initiated the SPADRAM project to identify means of reducing and mitigating signals passed at danger (SPADs). This followed the realisation that the introduction of a nation-wide Automatic Train Protection system (ATP) was not reasonably practicable (1).

The SPADRAM project presented it's finding early in 1995 and recommended the development and trial of a Drivers' Reminder Appliance (DRA) and a Train Protection and Warning System (TPWS). These recommendations were accepted by Railtrack and were subsequently included in the Train Protection Strategy that they presented to the Department of Transport in November 1995.

Compared to TPWS, which is an automatic system, DRA is a relatively simple manual system that requires trainborne equipment only. It is designed to stop drivers starting away against a signal at danger and provides a pushbutton that when activated illuminates red and prevents power from being taken.

Following the successful trial of DRA and the outcome of cost benefit analyses, Railtrack concluded that the installation of DRA was reasonably practicable. In 1996 Railtrack consequently issued a Railway Group Standard to mandate the fitting of DRA to all passenger driving cabs by 5th December 1998. Only fleets that were life expired or fitted with other safety systems such as tripcocks or trial ATP equipment were granted exemptions.

Using a single 'turnkey' supplier the Rolling Stock Leasing Companies (ROSCOs) successfully introduced this modification to over 4500 cabs, at over 40 sites within the mandated timescales. This was the first fleet-wide mandatory modification to be undertaken within the privatised railway and as such provides a useful model for future projects of a similar kind.

2 Development of TPWS

DRA was the first stage in the strategy for reducing SPADs. Whilst this modification was being implemented, Railtrack was developing, testing and trialing the next stage in the strategy, TPWS.

Compared to DRA that only addresses one kind of SPAD and is dependent on driver activation, TPWS offers significantly more safety benefits. The aim of the system is to both reduce the number of SPADs and the consequences should a SPAD occur by providing 'Trainstop' and 'Overspeed' functions. The 'Trainstop' function acts to automatically demand an emergency brake application should a train pass a signal at danger without authority. The 'Overspeed' function acts to automatically demand an emergency brake application if a train approaches a signal at danger at a speed likely to result in the signal being passed at danger.

The TPWS pilot trials commenced in November 1997 and are ongoing. The initial trials were undertaken using nine Class 319 units, leased by Thameslink, and ten signals in the Luton and Three Bridges areas respectively. The TPWS equipment (both trackside and trainborne) for the pilot trials was developed and manufactured by Redifon MEL.

Although the initial trials quickly proved that the equipment and the system could operate with a high degree of reliability, it remained questionable whether the fitment of TPWS was reasonably practicable and if so, to which signals and which trains. Whilst the industry debated these issues and various cost benefit analyses were undertaken, the Health and Safety Executive (HSE) intervened and drafted 'The Railway Safety Regulations 1999'(2).

These Regulations, eventually issued in July 1999, mandate the introduction of a train protection system (such as TPWS) that will cause the brakes of a train to apply automatically if the train:

a, passes without authority a stop signal such passing of which could cause the train to collide with another train, except a stop signal on the approach to an emergency crossover; and

b, travels at excessive speed on a relevant approach (an approach to a protected signal, speed restriction or buffer stop).

The deadline for the introduction of the system is effectively 1st January 2004 and after this date trains will not be permitted to operate unless a train protection system is fully operational.

3 Responsibilities for implementation

The responsibilities for meeting the mandatory requirements of the Regulations are detailed in Railway Group Standard GK/RT0090 'Requirements for a Train Protection and Warning System'(3). In accordance with this Standard the Infrastructure controller, Railtrack, is responsible for implementing and maintaining the TPWS trackside system and the Train Operators, are responsible for implementing and maintaining the TPWS trainborne system.

For passenger rolling stock, in accordance with their vehicle leases, the responsibility for implementing TPWS transfers from the Train Operating Companies (TOCs) to the three ROSCOs, Angel Train Contracts, Forward Trust Rail and Porterbrook Leasing Company.

The remainder of this paper will discuss how the ROSCOs intend to fulfil their obligation and implement the TPWS trainborne system to over 3500 driving cabs by 1st January 2004. Following a description of the current TPWS equipment, the paper will discuss the proposed implementation strategy and highlight project risks. Within these sections the interface between the trainborne and trackside equipment and the affect of TPWS on vehicle reliability will be discussed.

4 TPWS Functionality/Equipment

To provide the 'Trainstop' and 'Overspeed' functions, described earlier, both trackside and trainborne equipment is required. The performance of this equipment is defined within Railway Group Standard GK/RT0090 and at present only one manufacturer, Redifon MEL, supplies the equipment.

The trackside equipment consists of pairs of track-mounted loops controlled by electronic cabinets that are located at the side of the track. Each cabinet contains a power supply, a Signalling Interface Unit and a Trainstop or Overspeed Unit. When energised each loop transmits at one of six prescribed frequencies (f1 to f6), depending on their function ('Trainstop' or 'Overspeed'), the direction of travel and the position of each loop within the pair.

The trainborne equipment consists of a TPWS aerial, which detects the transmitted frequencies, a TPWS control unit, which interprets the transmissions (and if necessary demands a brake application) and TPWS visual and audible indicators. For each installation, additional brackets and wiring are also required.

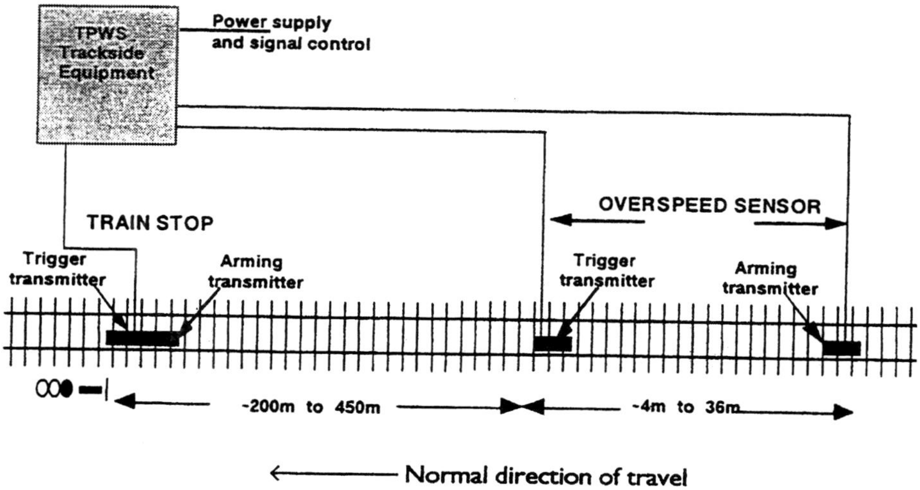

Figure 1. TPWS layout.

'Trainstop' protection is obtained by placing the transmitting loops adjacent to each other and within three metres of the signal. When the signal is at danger the loops are energised and each loop transmits their prescribed frequencies (f3 and f2). As the train passes over the first loop, the aerial detects the first frequency, the 'arming frequency'. Whilst still detecting the 'arming frequency' the aerial then detects the second frequency, the 'trigger frequency' and initiates an automatic emergency brake application.

'Overspeed' protection is obtained by placing the transmitting loops up to 36 metres apart and a distance of up to 450 metres from the signal. The exact positions are dictated by factors such as the overlap distance, the gradient, the braking characteristics of the vehicles on the route, the mix of traffic and the line-speed. When the signal is at danger the loops are energised and each loop transmits their prescribed frequencies (f1 and f2). As the train passes over the first loop, the aerial detects the 'arming frequency', and starts a timer. If the aerial detects the 'trigger frequency' before the timer reaches the trigger delay time, the TPWS control unit notes that the vehicle is travelling too fast and an automatic emergency brake application is provided.

5 Implementation Strategy

Until late 1998 when the HSE issued a draft copy of 'The Railway Safety Regulations 1999', it was unknown to what degree TPWS would be implemented and who would be responsible for the implementation. Following the issue of the draft Regulations these details were clarified and the ROSCOs, accepting that they may become responsible for fitting trainborne TPWS, started to consider implementation strategies.

Following a series of ROSCO meetings it was decided that, as with the DRA project, a joint ROSCO approach should be taken for the implementation of TPWS. In this way the structure of the project is simplified, a common approach is taken with each vehicle and each TOC and significant economies of scale can be realised in terms of both time and cost. In line with this decision a Joint ROSCO Working Group was formed to develop and implement a strategy to meet the requirements of 'The Railway Safety Regulations 1999'.

In January 1999 an initial review of the potential implementation strategies was undertaken and it was decided that a 'turnkey' approach would be taken. It was subsequently agreed that a single contract would be let to cover the implementation of TPWS to the ROSCOs' combined fleet of over 3500 cabs. In accordance with this 'turnkey contract' the supplier will be required to:

- Take full responsibility for the design, supply and installation of TPWS equipment in accordance with the requirements of 'The Railway Safety Regulations 1999' and all relevant Railway Group Standards.

- Produce common designs and modification procedures for each type of vehicle.

- Ensure that each design of TPWS equipment provides the same driver interface and system functionality.

- Ensure that the designs and their application will not reduce the reliability of any existing train systems, and where reasonably practicable will improve reliability.

- Provide documentation to enable the ROSCOs and TOCs to maintain and repair the TPWS equipment.

- Manage the implementation of the TPWS project using industry best practice and recognised Project Management techniques.

- Procure material that is fit for purpose and will interface and operate with all the existing equipment fitted to the vehicles.

- Install TPWS equipment to an approved programme, when vehicles are not required for Operational Service.

- Obtain any approvals required to enable each vehicle fitted with TPWS to return to Operational Service.

The decision to award a single 'turnkey' contract was taken to simplify the structure of the project and to reduce the overall project risk by removing commercial barriers, improving inter-disciplinary communication and clarifying responsibilities. In this way, the TPWS design is more likely to work 'out of the box', but if problems exist immediate feedback is given and the appropriate action can be taken without having to decide who is responsible for the problem.

The award of a 'turnkey' contract can also increase the project risk, however, as responsibility for all activities is transferred to a single organisation. The ROSCOs recognise this risk and intend to minimise it through a detailed competitive tendering process and the award of a contract that addresses the major project risks.

6 Project Risks

The risks associated with a project of this size and logistical complexity are considerable. In theory if the project fails and vehicles are not fitted by 1st January 2004, these vehicles will not be able to enter service, the rail network will be considerably disrupted and the ROSCOs will lose their lease revenue.

The major technical risks are associated with the TPWS equipment, the interface of this equipment with the vehicles and vehicle availability throughout the installation programme. In producing an Invitation To Tender for the 'turnkey' contract the ROSCOs have evaluated these risks and included clauses that seek to mitigate, remove or transfer them.

6.1 TPWS equipment

The risks associated with the TPWS equipment fall into two categories.

Firstly there are risks associated with the single source supply of the equipment, which at present is only produced by Redifon MEL. This dependency on one company to manufacture and support the use of the equipment in service is a major concern to the ROSCOs. It is also a difficult situation to address, as it will take a second manufacturer at least 12 months to develop and obtain approval for a similar product.

Having identified the risks associated with a sole supplier, the Joint ROSCO Working Group has sought to reduce both the likelihood and consequence of these risks by negotiating an agreement with Redifon MEL. In accordance with this agreement, should all the TPWS equipment be purchased from Redifon MEL, they will ensure that:

- Prices for the TPWS equipment and future spares are capped.
- Buffer stocks of one month's deliveries are kept at the manufacturing plant.
- Buffer stocks of key components are maintained.
- All major sub components are dual or multi-sourced.
- A second manufacturing plant will be capable of producing TPWS equipment.
- A Parent Company Guarantee is provided by Thompson-CSF
- Design and manufacturing data packs are lodged in an ESCROW agreement.

The second category of risk associated with the TPWS equipment concerns the functionality and reliability of the equipment itself. Since over 3500 cabs are to be fitted with TPWS, this is again of great concern to the ROSCOs.

The required performance/functionality of the trainborne TPWS equipment is contained within Railway Group Standard GK/RT0090. This standard describes how both the trackside and trainborne equipment must interface with each other and dictates high reliability and availability figures for each system. The reliability figures (in terms of Mean Time Between Failures) for the trainborne system are reproduced below in Table 1.

Failure Category	**Mean Time Between Failures (Hours)**
Failure of Train Stop System (TSS) or Overspeed Sensor System (OSS)	100,000
TSS or OSS operates when not demanded by track sub-system	42,000
TSS or OSS failure requiring equipment isolation	825,000

Table 1 MTBF for Train sub-system.

Based upon these figures the ROSCOs can take some comfort that vehicle reliability will not significantly worsen with the introduction of TPWS. The question to be asked, however, is does the current TPWS equipment (supplied by Redifon MEL) meet these high reliability requirements? Based upon reliability predictions to MIL-HDBK217F the answer is no, with the figures for two of the three failure categories a factor of two too low. The figures for the third category (TSS/OSS brake demand/indication when not required) are compliant.

This is not as bad as it first appears, however, as reliability predictions are inherently pessimistic and in-service experience often shows significant improvements. Unfortunately the current in-service experience of TPWS only amounts to 1.2 million miles or 30,000 hours, so even though only one failure has occurred (when a brake demand was provided when not required) no conclusions can be drawn. Fortunately Redifon MEL have considerable in-service experience (over 16 million hours) with a product that is similar to TPWS, Correct Side Door Enable. Based upon the fact that the overall in-service reliability of this product is twenty-five times greater than it's predicted reliability figures it can be fairly assumed that the TPWS equipment will satisfy the reliability requirements of GK/RT0090. Further in-service experience will show whether this assumption is correct.

6.2 Vehicle Interface

The second technical risk is associated with the interface of the TPWS equipment with the various types of equipment currently fitted to the vehicles. The risk is that the TPWS equipment, designed and tested principally on Class 319 units will not operate satisfactorily with other types of vehicle. This may introduce delays to the project and/or additional costs.

The ROSCOs are seeking to reduce these risks by inviting the potential turnkey suppliers to inspect the various classes of vehicle during the tendering process and by insisting that trial installations are undertaken on each class early in the project. In this way any issues will be identified at an early stage and both delays and costs will be minimised.

Interface problems will mainly be associated with the existing Automatic Warning System (AWS) equipment around which the TPWS equipment is installed. One such problem has already been identified involving the interface of the TPWS control unit, which replaces the AWS relay unit, with the AWS voltage converters that are fitted to a proportion of the vehicles.

The problem concerns the various types of AWS voltage converters and the tolerance of the voltage that they supply. The Redifon MEL equipment has been based upon the latest AWS Specification, 36, and has type approval for this configuration. Unfortunately approximately 30% of the existing voltage converters are to earlier AWS Specifications, 22 and 24. These converters produce wider voltage tolerances than the TPWS equipment is designed for and as such the TPWS control unit may indicate spurious faults and in extreme circumstances affect the operation of the AWS.

Although it is most likely that no unsafe mode will arise, it is envisaged that, where appropriate, new voltage converters to AWS Specification 36 will need to be fitted.

6.3 Vehicle Installations

The greatest risk to the project concerns the installation of the TPWS equipment and more specifically the availability of vehicles throughout the installation programme. This risk arises from the relative complexity of the TPWS installation, which will involve:

- The fitting of the TPWS aerial to the leading bogie or underframe (See Figure1).
- The installation of the TPWS control unit, control panel and isolation switch in the cab (See Figures 2 and 3).
- The wiring of the aerial to the control unit (this may involve lifting the vehicle)
- The wiring of the control unit, control panel and isolation switch into the brake system.
- The testing of both the installation and the TPWS functionality.

It is estimated that this modification will take 30 man-hours per cab. This makes the ROSCOs' requirement to fit vehicles when not required for operational service difficult to achieve. Careful planning and TOC co-operation will be required to identify opportunities to fit the equipment and staged fitting should be considered. Flexible programmes should also be developed to enable vehicles to be modified at their home depots during routine maintenance or at contractors receiving heavy overhauls.

To maximise the opportunity to fit TPWS the trainborne and trackside implementation programmes will be developed and executed independent of each other. This is because each implementation strategy is significantly different and a co-ordinated approach is likely to introduce unnecessary delays. Thus trainborne equipment will be introduced nation-wide, depending on vehicle availability, and trackside equipment will be introduced on a geographical basis.

Figure 2. TPWS aerial arrangement fitted to a Class 319 unit.

Figure 3. TPWS control unit fitted to a Class 319 unit.

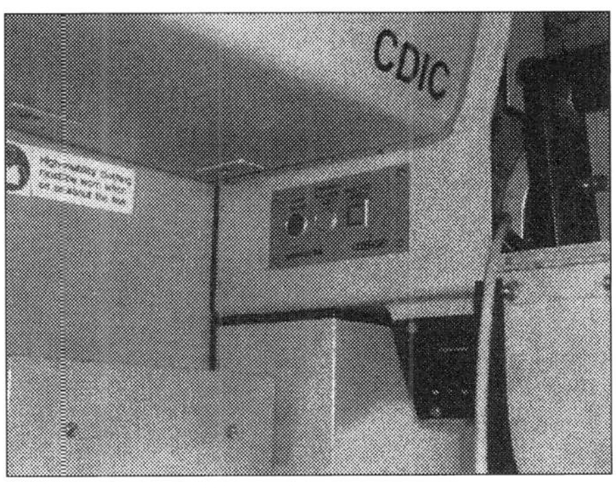

Figure 4. TPWS control panel fitted to a Class 319 unit.

7 The Benefits of TPWS

The safety benefits available from the implementation of TPWS will be only realised once both signals and trains on a particular line are modified. Close to the beginning of both the trackside and trainborne implementation programmes, therefore, the safety benefits will be minimal, as it is unlikely that a modified vehicle will interact with a modified signal. As the project progresses, however, the benefits will grow significantly and it is estimated that once all vehicles and all relevant signals are fitted TPWS will provide a saving of 1.7 equivalent lives per year (1). This conservative figure, which may actually be as high as 2 equivalent lives per year, equates to half of the annual SPAD related fatalities. Thus even though railways are the safest form of surface transport by a significant margin, through the introduction of TPWS it will become even safer.

The replacement of the AWS relay unit with the TPWS control unit provides additional benefits to the industry in that at a later stage the reliability of AWS can be increased. The replacement of the AWS relay unit itself offers some benefits, although the component is not inherently unreliable. It is understood that significant improvements in reliability can be made, however, through the subsequent replacement of the existing AWS receivers, with a poor performance record, with new lightweight receivers.

Whilst discussing reliability, it should be recognised that the introduction of TPWS and the subsequent disruption to the ageing AWS may also decrease reliability. To address this risk vehicle installations will be undertaken with great care and pre and post modification tests will be undertaken.

8 Conclusions

The implementation of TPWS will provide the railway with safety benefits in the order of 2 equivalent lives per year. Thus over the next four years, with the introduction of TPWS together with the introduction of new rolling stock and the disposal or modification of Mk 1 rolling stock, the safety of the railway will be greatly improved.

The goal for the industry, and in particular Railtrack, the TOCs and the ROSCOs is to ensure that the modification is implemented within the timescales and in a manner that minimises disruptions to services.

References

(1) "Train Protection & Warning System – Green Paper One", Safety and Standards Directorate, Railtrack PLC, February 1998.

(2) "The Railway Safety Regulations 1999", Health and Safety Executive, July 1999.

(3) "GK/RT 0090 - Requirements for a Train Protection and Warning System – Issue 1", Safety and Standards Directorate, Railtrack PLC, April 1999.

C570/006/99

Harmonising tilting trains with the infrastructure

C P BOOCOCK
Vehicle Acceptance, The QSS Group, Derby, UK

SYNOPSIS

The theory behind tilting trains is that faster curving speeds are possible on existing railway infrastructure. This paper illustrates that it is not sufficient to try to make tilting trains fit the infrastructure, nor is it sufficient to attempt to make the infrastructure fit the trains. There must be a total, "holistic" approach where both are treated as parts of a complete system. This paper takes a look at the interface issues in developing a reliable tilting train railway with a fault-free infrastructure.

1. WHY TILTING TRAINS?

1.1 High speed trains

The development of major high speed trunk railways in Europe and the far east has been in the form of specially aligned, new railway routes laid out for speeds of up to 300 or 350 km/h. Passenger trains for these routes have taken three specific forms:

- conventionally laid out stock on two bogies per coach, with traction distributed evenly throughout the train;
- power cars at each end with conventional carriages in between; or
- power cars at each end with the carriages articulated within the rake.

As soon as trains of these types are required to operate on other railways in Europe to reach destinations off the high speed lines they have to follow the rules of the routes on which they run. This reduces their running speeds to the line limits on these conventional railways, rarely more than 160 or 200 km/h, and well short of the potential for which the high speed trains were designed. Because conventional routes often include curves, these high speed trains have to reduce their speeds in consort with conventional trains as they pass through the curves, to prevent passengers experiencing discomfort from extreme lateral accelerations. Thus a German ICE running to Vienna and a French TGV heading for Milan are severely speed restricted for long stretches of their journeys.

As Europe has extended its high speed railway network, it has become increasingly difficult to justify the very high cost of new railways laid out for speeds in the region of 300 km/h. The high returns on investment that the early routes attained are simply not available on routes of lower strategic importance.

1.2 Return on investment
The case for tilting trains, whatever type of route they are designed for, is that they can shorten journey times on existing railway routes. Depending on the elasticities of the passenger markets on these routes, more people may be expected to travel. If business growth is sufficient to produce an adequate return on investment, tilting trains can have a future.

1.3 Tilt for secondary routes
As more railways have appreciated the benefits of higher speed travel, tilting technology has also spread. In Germany, for example, regional diesel multiple units operate several routes on curvaceous railway routes with useful journey time reductions, even though maximum speeds are no higher than with conventional trains. Many similar opportunities exist in the U.K.

2. THEORY OF TILTING

2.1 Safety and comfort
Lateral forces exerted by the wheels on the rails when traversing a curve at maximum (say, 6°) cant deficiency are still well within safety limits: a curving speed with 12° cant deficiency would still be nominally safe from derailment or overturning. The 6° cant deficiency limit is the accepted maximum at which passengers and train crew can tolerate the lateral force exerted on them from a comfort point of view. Indeed, there are those who counsel that 4.5° cant deficiency is more comfortable, particularly when track alignment defects are encountered.

2.2 Tilting the body
"Tilting" is probably the wrong word to use for what is done to the carriage body when the train runs through a curve. "Rotating" would be better. The carriage body is rotated a few angular degrees relative to the track around a point near the body's centre of gravity, such that passengers feel a much reduced lateral curving force. A body tilt of 8° enables trains to curve up to 30% faster than conventional trains. To rotate a carriage body and yet stay within the loading gauge necessitates narrowing the body profile above the waist, sufficiently to ensure that the body is within the designed kinematic envelope in all conditions, including hard tilt failure (i.e. when the body tilts the wrong way).

2.2.1 Passive tilt

The Spanish and Japanese build trains that use the lateral force on curves to swing the carriage body outwards on what is in effect a pendulum suspension device. In the Spanish Talgo trains, the carriage suspension is located at roof level, and the out-of-balance force on curves acts above the body centre of gravity. This increases the train's potential overturning moment. The angle of body tilt is thus restricted to about 3°, which limits the extra curving speed. This small passive tilt angle would render tilting uneconomic in many applications. Its main attribute is the resultant improvement in passenger comfort, rather than significant increases in curving speeds.

Another disadvantage of passive tilt is that the train has already to be traversing a curve before the body responds by swinging outwards. There is therefore a time lag during which passengers are exposed to a greater lateral force. At the limited incremental speeds normally run by these trains, this normally passes unnoticed by most passengers.

A major advantage is that the passive tilting mechanism is technically simple and reliable. Nonetheless, the rest of this paper deals exclusively with trains employing active body tilt, as it is these that push the technical boundaries of trains and infrastructure and which therefore are more demanding of engineers and operators.

2.2.2 Active tilt

Body tilt angles as high as 8° require positive, forcible displacement of the body, responding very quickly to changes in track curvature. In principle, there is a simple geometric suspension arrangement that connects the body and bogie through mechanisms able to rotate the body around a point close to its centre of gravity. The suspension elements are powered by hydraulic or electric thrust devices that respond speedily to track curvature changes. When maximum cant deficiency is detected, body tilt angle is at its maximum, the angle being chosen so as to compensate the passenger for up to 70% of the unbalanced lateral force.

One key element of the design is to select an appropriate speed of response for the tilt mechanism. With too slow a rate of change of tilt angle the passenger may feel the lateral force at discomfort levels. With too fast a rate of response, a standing or walking passenger could fall over. The choice of response is a compromise. Figures for the rate of change of tilt angle between 5° and 10° per second are commonly selected in modern train designs.

2.2.3 When to tilt, and by how much?

There are basically three ways of determining when a railway vehicle should begin to tilt:

- by means of a lateral accelerometer and a gyroscope, usually mounted on the leading bogie or on the vehicle in front. From these, the tilt mechanism can distinguish between a track irregularity and a genuine curve, and respond accordingly. Simplistically, the accelerometer output determines the angle of tilt, and the gyroscope decides if this is in fact a real curve;
- by means of transponders on the track bed to indicate the beginning and end of curves through which the train must tilt, coupled with on-board processing of speed and memorised curve data to calculate the tilt angle; or

- by means of on-board memorised route data coupled with on-board processing of curve and speed data to calculate the tilt angle required, possibly supplemented by satellite positioning equipment.

Of the above ways of determining tilt angle, the on-board accelerometer and gyroscope method is the most common, as it is not dependent on infrastructure-related or memorised systems, and in effect gives the train a live, "go-anywhere" capability, at least in theory.

3. TILTING IN PRACTICE

3.1 First essay

The first attempt at applying body tilt to a train designed for higher speeds on existing infrastructure was British Rail's Advanced Passenger Train (APT) which successfully ran on test at up to 256 km/h on parts of the West Coast Main Line.

The APT tilt system completely compensated for the lateral curving force felt by passengers. In practice, many passengers found this disoriented their sense of balance and the concept of tilt nausea emerged. Modern tilting trains only compensate for between 60% and 70% of the lateral force, and tilt nausea *per se* is said no longer to be an issue.

Figure 1: British Rail Advanced Passenger Train test formation about to head north from Quintinshill on a run that achieved 250 km/h in Scotland in April 1980.

The most significant change made to the infrastructure of the West Coast Main Line to accommodate APT was to back up the signalling system so that the trains could run safely at speeds that required longer stopping distances than did trains stopping from conventional speeds. Transponders were fixed at intervals on the sleepers between the rails on the whole route from London Euston to Glasgow Central. The transponders were "read" by the train as it passed over them. They relayed to the driver the speed at which the train needed to run to cope with forthcoming signal aspects and track speed restrictions.

3.2 Continental practice

Both the Swedes and Italians have railway main lines that cross mountainous regions and that thread deep valleys with curvaceous route profiles. Italy has projected its *Cisalpino* tilting trains across the Alps into Switzerland and France, and Switzerland is seizing tilting train opportunities in its own mountainous territory. European countries have developed tilting mechanisms that build directly on the APT experience.

In Germany there was a need to speed up trains on inter-regional routes, many of which have long curvaceous sections. That country has pioneered the application of body tilting on relatively mundane trains for more local applications. The principles, though, are the same.

Figure 2: Italian ETR460 high speed tilting train (Pendolino) approaches Firenze en route from Roma to Bolzano.

Figure 3: A tilting d.m.u.! German class 611 units arrive at Köln Hbf. on a secondary working from Saarbrücken.

Detail differences have emerged among the many designs now in existence. Some use a bolster between the vehicle body and bogie on which to mount the tilt suspension elements. Others put the tilt linkages below the bolster and the vehicle suspension above. The Japanese sometimes mount vehicle bodies on rollers on curved tracks on top of the secondary suspension, in place of geometric linkages, a system that has been exported recently to Australia. While hydraulic or electric tilt rams are the norm, there are some trains that use differential inflation pressures on vehicle springs to induce body tilt.

3.3 Characteristics of tilting trains

Tilting trains respond to inputs from the infrastructure in much the same way as conventional trains, but because their speeds are higher the inputs are disproportionately greater. The key element is the need for tilting trains to recognise an approaching curve. If the curve's transition is as designed, it should lead the tilt mechanism to begin its work sufficiently early to ensure the carriage body tilts progressively as the curve is entered.

Track irregularities have the potential to cause the tilt mechanism to "mis-read" the entry to a curve, either by acting as though there is a curve when there is none, or by delaying tilt when a real curve is entered. Both effects are uncomfortable for passengers and train crew. Tilting train manufacturers claim that their mechanisms are designed to detect and ignore track irregularities, but this author has experienced one or two very odd movements on tilting trains at locations on the continent where track alignment is not up to scratch. On the other hand, the ride and tilt response on good track can be excellent.

3.4 Secondary routes

One should emphasise at this point that the journey time benefits of employing tilting trains are just as good, if not better, on medium-speed routes with many sharp curves as they are on high speed routes. Any route on which trains routinely have to brake to traverse frequent curves and then re-accelerate after them should be studied to see what opportunities tilting trains offer.

4. INTERFACES WITH THE INFRASTRUCTURE

4.1 Leave the infrastructure alone?

Let us first explore the potential for designing a tilting train to fit the existing infrastructure on a typical route without modification, be it a main or secondary line. Is there any potential for not doing anything to the infrastructure and yet produce due economic betterment to the operator? Clearly, to do this a tilting train must fit the infrastructure in all respects. It must fit the loading gauge, which we can design it to do. It must be within the route's track force limits and axle loadings.

However, tilting trains may need to slow down at places on an existing, unmodified railway if any of the following apply:

- signal spacings are too short for stopping from higher speeds;
- track workers do not have refuges or approach-operated automatic warnings;
- there are level crossings of various kinds on the route;
- electrification power supply is insufficient to support higher speed operation;
- overhead contact wires are insufficiently tensioned for higher speed operation; or
- track maintenance, particularly on curves, is not to a higher speed standard.

So, if we try to zero the need to alter the infrastructure to operate tilting trains on it, their use will not bring all the benefits proposed. Thus there has to be a compromise.

4.2 Make the infrastructure fit the train?

This is tantamount to designing an ideal train and then offering it to a railway and finding that the infrastructure has to be expensively modified for it to run. [This has happened before - in Germany, some tunnels needed internal surgery at about train cantrail level to let the class 610 tilting d.m.u.s safely through.] Other propositions under this sub-heading are really so preposterous as to need no further attention! For example, each of the interfaces listed in paragraph 4.3 has a price on it, either to make the train fit the infrastructure or the infrastructure to fit the train. In almost every case it makes no practical, economic or financial sense to change the infrastructure to fit the train where there is a reasonable choice of making the train fit the infrastructure instead.

Section 5, however, deals with those important issues where unavoidable changes must be made to the infrastructure if tilting trains are to operate correctly and safely.

4.3 Compromise - a train-to-infrastructure interface specification

The reality is that there has to be compromise. An operator may be lucky enough to find a route for a tilting train fleet on which the infrastructure does not need a great deal of expenditure. APT and the West Coast Main Line in its early 1980s condition were a case in point, though even there the cost of fitting signalling transponders over 647 km of double track route was significant!

It was for reasons such as those described earlier in this paper that Railtrack demanded a statement of the interface limits of trains and infrastructure when it was considering its part in the development of the West Coast Main Line for the new Virgin high speed tilting trains. This statement was known as the interface specification (or "capital envelope" because any infringement of its agreed boundaries would result in extra infrastructure expenditure to be negotiated with the franchisee). Such an interface specification needs to be agreed for any proposed application of tilting trains to any route, whether main or secondary line.

An interface specification should address, as a minimum:

- the maximum track cant on the route;
- the maximum permitted cant deficiency on the route;
- the maximum permitted angle of tilt of the train (this must take into account any additional degrees tilt caused by rolling of the body on its suspension);
- the percentage incremental speed permitted for tilting trains traversing curves;
- the maximum rate of change of tilt;
- the kinematic envelope within which the train, including its pantographs, must conform;
- the policy for handling within the kinematic envelope the effects of tilt failure at speed, including reverse tilt;
- the design of curve transitions on the route;
- the deviations permitted from curve transition design before corrective maintenance is needed;
- limits on track irregularities;
- the relationship between signal spacings and tilting train braking performance;
- the method for signalling the tilting trains, if different from other trains on the route;
- the method for identifying differential speed restrictions to train drivers;
- the method for overriding driver-controlled speeds if safety is at risk; and
- threshold speeds above which changes to track safety rules must apply (for such as level crossings, turnouts on curves, safety refuges and track worker warning systems).

The infrastructure owner may well wish to add to this list such features as maximum permitted axle loads; maximum track forces, both lateral and vertical; enhanced electrification power supply; overhead line re-tensioning; means of detecting failing axle bearings; emergency equipment requirements; failed train rescue arrangements, and so on, but these are features which surely should be addressed when any new trains are contemplated, tilting or not, so are less relevant in this paper.

5. INFRASTRUCTURE FOR TILTING TRAINS

5.1 Track

5.1.1 Track alignment

On a railway where the track is aligned in accordance with scientifically-derived standards and where all curves have transitions that follow established rules there should be no need to alter the track when confronted with a fleet of tilting trains. Such was the case with the West Coast Main Line and APT, and is the case now in many of the European countries contemplating tilting train operation.

Nonetheless, track alignment cannot always be perfect. It is usually sufficient to ensure that track misalignments are corrected at locations where conventional carriages with passive suspensions habitually fully compress their lateral bump stops. For tilting trains, such misalignments should normally be taken in their stride, but would the wave-form variations in top level that have appeared on some main lines in the U.K. be a potential source of confusion to tilt mechanisms?

One must also consider the effect of rail surface irregularities, such as corrugations, on the inherent transmissibility of noise through the train's tilt mechanism into the carriage. A recent ride on a Pendolino that had wheel tread damage was unpleasant in that much of the noise came into the passenger compartment through the body frame. There is a message for train designers here - how can the tilt mechanism be isolated from potential noise paths?

5.1.2 Track maintenance

One area of track maintenance that that requires particular care is the alignment of curve transitions, because the majority of tilting train designs use these transitions to trigger the tilting mechanism. The Norwegian Railways (NSB) have made a detailed study of how to overcome their traditionally short curve transitions so that future operation of tilting trains can be effective. NSB have concluded that not only does the design of transitions need to be optimised for the tilting trains, but maintenance of transitions must also be permanently improved. Thus, as experience is gathered in the future in the U.K. it may well prove to be necessary to specify tighter tolerances on transitions here also.

Because tilting trains can go faster over specific stretches of a route, track misalignments that were acceptable at lower speeds may no longer be so. General track maintenance tolerances may therefore need to be tightened up, at least to the level that conventional trains would require if they were to be operating at the same enhanced speeds without bump stop contact.

5.2 Signalling and train control

On some secondary routes it may well be possible to operate tilting trains by specifying higher braking rates that enable the trains to stop within existing signalling stopping distances. On routes where tilting trains are introduced at the same time as a general increase in maximum line speed it is unlikely that existing signal spacings or sequences will be adequate. Relocating signals is expensive. Additional signal aspects, or flashing aspects, on existing signals may be a cheaper solution.

Other forms of control such as cab signalling are sometimes appropriate because different speed messages can be transmitted to different types of train, allowing for differential speeds on curves, for example. These systems might include:

- train protection warning system (TPWS);
- automatic train protection (ATP);
- transmission-based signalling.

A tilting train that traverses a curve, say, 30% faster than conventional trains is operating much closer to its overturning speed, though not so close as to be unsafe unless the driver permits speed to exceed that intended. Nonetheless, the experience in Italy of the Piacenza accident, where an ETR460 tilting train overturned when passing through curves at excessive speed, indicates that there is considerable wisdom in insisting on a form of automatic overspeed override equipment to be standard on tilting trains. Railtrack's TPWS system could be a minimum form of overspeed control for those routes not expected to be equipped with transmission based signalling or other forms of automatic train protection.

5.3 Lineside changes

5.3.1 Level crossings

In the U.K., different rules apply on railways with speeds over 160 km/h than below. There is another threshold above 200km/h. So, for example, when higher curving speeds on a main line enable a tilting train to break the 160 km/h or 200km/h barrier on certain curves, level crossings on those sections may need to be upgraded, or even replaced by bridges or underpasses.

5.3.2 Safety on the track

The safety of track workers also depends on the approach speeds of trains. Again, there are different rules for different speed bands. Tilting trains will break these speed barriers on certain curves resulting, for example, in the need to provide track-side refuges where there were none before, or fixed warning systems to protect staff where look-outs were appointed before.

6. MAKING THE SYSTEM RELIABLE

6.1 Avoiding complication

Much of what this paper has discussed involves adding potentially complex technical equipment to trains and to signalling systems. Greater complexity often leads to lower reliability, because there is more to go wrong. For example, if a tilt system fails the train speed may need to be lowered to that of conventional trains, thus losing time on the journey.

Modern signalling systems, too, threaten to be complicated, particularly if backed up by safety support systems such as ATP (Automatic Train Protection). A composite system such as transmission-based signalling that gives required speed inputs to the train can undertake both the role of the signaller and the role of the safety protector, in theory with relative simplicity.

6.2 Redundancy

Smaller devices on the train tilt systems such as accelerometers and gyroscopes can be duplicated to gain reliability from redundancy. Microprocessors can be set in multiple, to operate when there is agreement between them.

6.3 Self-contained tilt control

The most accurate forms of tilt initiation are clearly those incorporated on the vehicles themselves. These translate the lateral and rotational forces that are input from the track, to the instructions to the tilt mechanism to tilt the body at the required angle and tilt speed. There are two inherent disadvantages of such a system:

- the leading vehicle receives its tilt instruction a little later than ideal, so that the tilt is operated slightly late; recent experience suggests that passenger discomfort from this delay is marginal;
- the multiplicity of mechanisms throughout the train carries a reliability risk which has to be minimised by design.

6.4 Externally-initiated tilt control

In theory it is possible to initiate tilt from inputs external to the train. These possibilities include:

- transponders on the track specifying the position of the start of a curve transition - this is not ideal, as the start of tilt may sometimes need a speed reference as well;
- global satellite positioning to indicate the start of a curve - current technology in this field is not accurate enough to differentiate between adjacent tracks, nor to be precise about the right point at which to initiate tilting.

Neither of these systems offers the same accuracy of control nor the "go anywhere" potential as do on-board tilt control systems.

6.5 Mature technology

This paper has shown that tilting trains are not new. Train tilt technology has grown over 20 years, and indeed is more mature than that of some of the signalling and train control systems now being proposed to support it. This maturity will surely stand the new trains in good stead as tilting trains become the common solution to the demand for shorter journey times.

C570/003/99

Maintenance depots – can they ever be fault-free infrastructure?

P S M RAMSEY
Parkman Limited, Nantwich, UK

SYNOPSIS/ABSTRACT

The development and maintenance of the infrastructure of depots themselves have perhaps been low on a train operating company's priorities. This paper shows how important is the correct development of depots, and how crucial is the ongoing maintenance of depots themselves in providing the quality of service that the train operators and their passengers want. The maintenance of depots is considered alongside other similar activities, in the public and private sector, from which lessons can be learnt for better depot infrastructure maintenance in the future. The contractual and operational and control arrangements needed for best results and most reliable operation are explored, together with examples of the way these bring benefits. Possible future arrangements will also be explored.

INTRODUCTION

What do we consider when we think about a depot? Do we think of it only as an on-cost to our business? Or is it an asset, in every sense of the word, which enhances the provision of the prime assets, the trains? Does the depot add value to the servicing of the trains we give? These and many other issues will be explored in this paper, which considers the whole ethos of depots, and then examines ways in which the maintenance of the depots themselves is carried out.

DEVELOPMENT OF DEPOTS

Depots have grown from being just a location where locomotives could be stabled and maintained, to the multifleet depots which we see to-day. Carriage maintenance has previously been carried out in open yards, with only very rudimentary examination and maintenance possible. Freight vehicle maintenance has been even more primitive, with open sidings being the norm for many generations. The concept of maintenance of vehicles under cover used to be for locomotives only. Carriage sheds were little more than a cover for coaches - little maintenance was needed and little was carried out. This has now altered so that almost all vehicles are maintained under cover. The new depots for High Speed Trains in the early 1970's brought about a change in practice, which has spread to whole trainsets being maintained undercover at once. Wagons, too, have now their planned maintenance in bespoke depots. A recent example is the new two-road 850m long maintenance depot for Eurotunnel at Coquelles in north western France which can take, complete, two Eurotunnel shuttles at once. Perhaps the most spectacular is the North Pole depot for Eurostar, which is spread over some 4 km of land with the various facilities all available.

OBJECTIVES OF DEPOTS

What then are the objectives of a depot? Its main aim must be to enable all trains to be maintained so that they are available for use as planned. Depots should help this process. Depots should be equipped so that they can deal quickly with any foreseeable requirement for maintenance. Depots themselves should be so built and maintained that they need low maintenance themselves.

LOCATION OF DEPOTS

Where are depots best placed? Ideally where there is little empty stock running needed. In this an example is Morden on the Northern Line of London Underground. Another is the depot for the Line 3 of Milan Metro where the line finishes in the depot. However geography can tend to militate against this. We have a multitude of depots which are some distance from terminal stations, because land was not otherwise available, and which incur empty train running Wembley and Oxley at Wolverhampton are examples. We have some which are adjacent to the line of route but at an intermediate point, usually one which is a starting point for at least some of the train service which is operated from there. These, too, can avoid empty train running by trains running directly from the depot into service. Such is Golder's Green on the Northern Line of London Underground.

DEPOT INFRASTRUCTURE

What infrastructure does a typical depot have? There are now buildings in which the day-to-day maintenance of trains takes place. There are also sufficient sidings for stabling trains ready for use. The buildings are long enough to take train-sets so that there is no need for uncoupling the individual vehicles. This is particularly true if the sets are articulated, or maintenance downtime is limited. The buildings therefore have tracks inside, and depending on the maintenance planned, will have pits and platforms at various levels for access in the best possible way to all those items which require inspection and attention. The pits may not just be between the rails, but may extend in steps either side of the railway tracks. This of course depends on the design of the trains going to be maintained, as well.

To meet the objectives of a depot, the infrastructure needs to be efficient and fault-free. If you consider the way in which the service each day depends on the output of the depot, then the infrastructure must be of a high standard itself. The track needs to be comparable with that installed and maintained on the main line. Depots ideally need two access points into the main line, one for use day-in, day-out, the other for use occasionally and in an emergency. Whether the depot needs to be signalled and to have power operated points is a matter of safety and economics. On a long depot where separate parts interface, the answer must be yes. If there are only drivers as traincrew, then it is almost essential to have powered points and signalling. It is safer to have as few people as possible having to walk round stabling sidings. The structures of the buildings need to be substantial, themselves requiring least maintenance, and also these days as insulated as possible, to avoid the extremes of winter and summer. The types of vehicles being maintained need to be considered in relation to ventilation. Where there are to be diesel-engined vehicles, some form of extraction needs to be provided, ideally with assisted extraction, but not such as to leak rainwater backwards into the depot. I have already mentioned the actual areas of maintenance themselves. These need careful design from the start. New depots give a super opportunity for the floor layout to be the best for people to work on. The example of the Dutch railways in the late 1980's where they experimented with different heights of floors and pits led to the ideal layout - for their staff and their electric multiple unit fleet - at their depot at Onnen.

The use of standard components in the track layout helps in maintenance, and what we are trying to avoid by good design, derailments.

The topic of the maintenance of equipment at depots raises the question about the actual equipment that a depot requires. If the depot is to act as a servicing point as well, then there needs to be somewhere specifically designed, or it is taken account of in the design of the facilities so as not to have any conflict, of e.g. rubbish lying where maintenance is being carried out. Carriage washing plants which work consistently are vital in the new railway. The optimum types of flails and brushes match the profile of the vehicles, as well as their front or back ends for which equipment has now been developed, and the best practice needs to be spread throughout the industry. Washing plants have often been a continuous problem in respect to unreliability. The reputation of a train operating company can be very much influenced by the cleanliness of its vehicles. There is an opportunity to make sure that the equipment is very reliable, and that means that it can also be used in winter. Washing plants inside heated buildings may be the solution.

New too in the last few years are the toilet effluent points. Sufficient of these need to be available for trains to be quickly dealt with. The pitch of these points needs to match the vehicles, not only in coupling type, but also in combination of train sets. Like other hoses and cables, their storage and handling need to be designed. Otherwise there is excessive cost through damage and replacement of these vital components. It is also vital that all this equipment remains fault-free, as failure to be so can seriously disrupt the maintenance programme and operating practice of the train operator

Depot operators need to make sure they have sufficient pieces of equipment to cover most eventualities, for instance battery chargers. Where these are placed and how many are required form part of the planning for maintenance of a fleet. It is certainly the case that the design of the vehicles needs to be matched with the facilities. A low voltage charging system throughout the depot may be better than many individual chargers which are subject to damage as they are dragged round the depot. But the vehicles must be able to accept these low voltage connections. Alternative systems for connecting up the high voltage points on vehicles are an alternative, and these can then supply heating and battery charging within the vehicles of a train set.

As in many other spheres, the simpler the equipment is the better. For instance, the generally specified standard for depots is for a ground wheel lathe. For a two axle machine, there will be little change out of £5M to install of these ready to use. And there is the running cost of it. In terms of availability of trains, if a wheelset needs to have attention, it is better to have a wheel/bogie drop, with a replacement bogie available, and to swap these rather than reprofile wheels in situ. The wheels can then be sent away to an above floor conventional lathe for minimum material to be taken off, not a feature of underfloor lathes.

The floors of the depot need to be easy to maintain, and with non-slip surfaces. Certainly depots with working areas in red, pathways in green are attractive, but they do need constant attention. Plain concrete may well be sufficient, but with pathways delineated by white lines. The depot design is important, and maintainability needs to be part of the actual design. Architects do get awards for maintenance depots that look aesthetically pleasing, but do not have to live with the task of changing a lightbulb in the ceiling which means that you have to shut part of the depot and avoid the smoke detectors in the roof detecting the presence of someone, whilst specialist lifting access is hired to restore lighting levels to that planned.

Safety is always part of the depot, as indeed is the competence of the people, not only those who maintain the trains, but those who shunt and drive the trains around the depot, and those who maintain the infrastructure itself. Nowadays with the complexity of the depot we have seen, it is impossible for any one person to know all there is about all items of infrastructure on the depot. Some of those involved in maintenance of the infrastructure have their own expertise, and need to have their own competence enhanced by training and assessment of competence. Maintenance of equipment for its major overhauls may be by the Original Equipment Manufacturer, or a service organisation which specialises in particular pieces of equipment.

Access, trespass and vandalism is always of concern to those at a depot. Not only do we not want to see any youngsters hurt or even killed by being illegally on the depot, but we want to be able to turn out our trains in pristine condition without being pride of the local graffiti artists. Substantial fencing is now vital round a depot, and sufficiently far either side on the main line to deter anyone from coming along it and in by the rail access. And new ways of people entering depots such as through turnstiles and with security cameras are now the norm. Swipe cards are now sufficiently developed to be reliable. Entry for those legitimately coming on to the depot must be easy, but for those with evil intent must be made very difficult.

ASSET MANAGEMENT

All the assets of a depot need to be known and logged. This is not only from the point of view of those who are maintaining trains, but it is also to make sure that these depot assets give their best availability, as well as identifying their maintenance and giving them their statutory inspections etc.

EXAMPLES AND LESSONS FROM OTHER RELATED SPHERES

What can we learn from other industries relative to the maintenance of depot infrastructure? Examples are given below of types of infrastructure which also need maintenance, and from which some lessons may be learnt.

1 Motorways

Motorways represent another mode of transport, and the way that they are maintained is useful to examine. The Highways Agency has overall responsibility, and they have contracted this out to Civil Engineering consultancies to manage the renewal and maintenance. Nowadays it is not only the road substructure and surface that needs attention, but such aspects as roadside telephones, gantry and dot-matrix signals and protective barriers. The arrangements for possessions need to be just as precise as those on railways, and the safety of those involved in preparing for roadworks and involved in carrying them out is paramount. There is also the need to have provision to deal with emergencies, not only those which the police, ambulance and fire brigade can deal with, but also with all sorts of contractors who can deal with the mangled mess after accidents, and the environmental companies to deal with, say, 10,000 gallons of fuel oil or milk.

The arrangements for maintenance are that the Highways Agency, as the client, appoints a consultancy as its Managing Agent for dealing on the Highway Agency's behalf with routine maintenance and maintenance renewals contracts. (For major new work, the Client, the Highways Agency, again appoints a consultant as the Design Consultant as their managing agent, and the contractors as well. The Highways Agency is the informed client, and certainly still stays in touch.) The professional expertise lies with the Managing Agent, who has no axe to grind, has the interests of providing best value for money from the budgets, yet is one step removed from the contractors. The Managing Agent is driven by standards and codes of practice, and best value, over the long term. The Managing Agent can have the holistic approach to the whole activity, long term. The client remains in control because he gets straight feedback from the Managing Agent. Interestingly for access for possessions to the motorway, the Managing Agent arranges and agrees this, with the maximum benefit to users and client the prime consideration. All the statutory bodies - the police, the other emergency services, the local authorities - are dealt with by the Managing Agent to make the arrangements for possessions, and he also has the role to inspect.

The Managing Agent is in a very good position to review standards and to investigate and recommend changes to improve. Thus the new motorway information signals have come about by one managing agent thinking through how better to install and then maintain these signals, with the result that the installation has been much simplified, and access can now be gained, and then maintenance can be carried out internally, without any disruption to traffic at all.

2 The Fire Service

Fire Stations have buildings and plant and equipment which are similar to those with which railway engineers are familiar, though an exception is perhaps the escape pole! In the instance considered here, the heavy maintenance of the Fire Stations, the control offices and the surrounding areas is managed under a five year term commission to provide property management and construction related professional services in a partnership contract. In the particular Fire Service there is a portfolio of 41 fire stations, ancillary buildings and offices. The portfolio exceeds 50,000m^2. The scope of works includes redecoration, repairs to fabric and boundaries, yard resurfacing, roof replacement, renewal of appliance bay doors, floor tile renewal in appliance bays, plant renewal in Fire Services Headquarters and construction of a live fire training module. All this is work carried out on 'live' fire stations, and with the control centre having to be totally available, with no break in communication facilities, 24 hours a day, all the time. By means of working in partnership with the Fire Authority, the money available is able to be identified for priority work. Surveys of the assets show what needs to be done to keep these assets operational. Rolling programmes for 5 and 10 years are able to be achieved so that the Fire Service is able to meet its performance targets for dealing with their emergency calls. The result has been that the Fire Service knows what maintenance is going to cost, that it is such that it will keep the Fire Stations in a fully maintained condition, and confidence is built up both of the Fire service staff and the contractors that the facilities are always available. It is suggested that this type of contractual relationship can be transferred across into the rail industry.

3 Local Authorities

Local authorities are also seeing the benefit of a different style of approach in the maintenance of their infrastructure. A Borough Council has entrusted the maintenance of their building population to Civil Engineering consultants. In now a second 5 year commission, these consultants were appointed to provide surveying, engineering and property management advice on all local authority public buildings, nursery and primary schools. What this has done is to put the management of the facilities on a firm footing, with improving levels of service, making use within 13 specialist trade sections of a group of contractors which is providing best value for money at all times. Some 30,000 individual repair requests are dealt with each year. By focusing on the real needs of the customer, the engineering expertise of the Civil Engineering consultants has been able to significantly improve the performance of the Local Authority in this area which so often is regarded as low in priority.

For another local authority, the management of their stock of bridges is covered by a managing agent consultancy. Not only is the inspection of the bridges undertaken, but the arrangements for the contracts are fully managed on behalf of the client.

CONTRACTUAL AND OPERATIONAL ARRANGEMENTS

What contractual and operational arrangements are seen to work best with maintenance of the depots? We recall the three standard forms of contract which Train Operating Companies were given on their sale in the mid 90's, with the former Infrastructure Maintenance

Companies. These covered Buildings Maintenance, Plant and Equipment Maintenance and Engineering Services - Infrastructure: really, Buildings, Plant and Track/Signalling/Overhead Line. These were put together quickly to try and mirror a non too satisfactory situation. That the work and the value of it were naturally distributed among each of the local Infrastructure Maintenance Companies is a matter of history. And those companies were also sold with the maintenance of the mainline network locally too. So the previous practice of concentrating on their main contract, the main line, naturally took precedence. And the contracts were in themselves a reflection of a conflict/ adversarial stance between the two parties. It is now seen that rather than an adversarial position, we need to get away from conflict, and together try to get the depot to run smoothly. The depot operator needs to rely on professional expertise, on the contractor and his staff. The staff need to be well trained. There needs to be a share in success of depots.

The management of contracts is vital to the success of having fault free infrastructure. There are several model options now available. One can be for the client - the depot operator - to just manage the contractors themselves. But this begs the question of the professional expertise which this client can be expected to have. Rather it is suggested that a Managing Agent with the right professional expertise is commissioned by the client to act on his behalf. This can be in between the client and the contractors, to act as a manager on the client's behalf, as well as a facilitator to the better outcome. Alternatively, the client can have consultancy/management support in reviewing the arrangements for maintenance, with the contracts managed by the client directly. A third way is for the client to appoint a maintenance management partner to act as a facilitator to developing a partnership solution with the contractors and, in this way, to develop the relationship for the long term future. Certainly partnership by one model or another is seen to bring benefits to all parties.

The ability of the maintainer to do planned maintenance and to respond to emergencies is crucial to the smooth operation of the depot. Some emergencies can be covered by having spare pieces of equipment always available. Others have to be dealt with when they occur. The ability of the arrangements for a maintainer to respond to an emergency because his people live locally is a key. Although the competence of his people to maintain equipment to avoid this is also. A relationship of trust needs to be developed between the professional expertise, the maintainer and the depot operator, and they need to share in the success of the depot.

CONCLUSION

In this paper we first of all looked at the infrastructure now required at a modern depot, and pointed out some of the aspects which need to be considered right from the early stages of development. Depots need to match the trains to make for the best facility, and need to be able to efficiently deal with the train maintenance required. The maintenance of the infrastructure of the depots themselves needs to be such that it enables the maintenance of trains to be trouble-free.

In the examples of the motorways, the fire stations and the Local Authority infrastructure we have seen what can be done to address similar activities in a different way, in particular within partnerships.

That there are no delays to trains out of a depot, and that the full service is provided day-in, day-out is a measure of the success of a Train Operator. Whilst the design and maintenance of the trains themselves is vital to achieve this, it is also a measure of the success of the arrangements for the design and maintenance of the depot facilities themselves. These arrangements reflect the relationship between the depot operator and his depot maintenance regime. What has been tried to be shown in this paper is that this is able to be accomplished by a partnership between the depot operator and the maintenance contractor, and is able to be improved when facilitated by a consultancy with professional expertise in maintenance of assets of all kinds. Together they will enable the depot infrastructure to be fault free.

Authors' Index